SPRINGER LABOR MANUAL

Springer

Berlin
Heidelberg
New York
Barcelona
Budapest
Hong Kong
London
Milan
Paris
Santa Clara
Singapore
Tokyo

Th. Köhler D. Laßner A.-K. Rost
B. Thamm B. Pustowoit H. Remke

Quantitation of mRNA by Polymerase Chain Reaction

Nonradioactive PCR Methods

With 15 Figures

Springer

Dr. rer. nat. Thomas Köhler
University of Leipzig
Department of Medicine
Institute of Clinical Chemistry
and Pathological Biochemistry
Molecular Biological Laboratory
Paul-List-Straße 13/15
D-04103 Leipzig

Dr. rer. nat. Barbara Thamm
University of Leipzig
Department of Medicine
Institute of Human Genetics
Research Laboratory
Philipp-Rosenthal-Straße 55
D-04103 Leipzig

Dirk Laßner
Prof. Dr. med. Harald Remke
Anne-Katrin Rost
University of Leipzig
Department of Medicine
Institute of Clinical Chemistry
and Pathological Biochemistry
Molecular Biological Laboratory
Liebigstraße 16
D-04103 Leipzig

PD Dr. habil. Barbara Pustowoit
University of Leipzig
Department of Medicine
Institute of Virology
Liebigstraße 24
D-04103 Leipzig

Die Deutsche Bibliothek - CIP-Einheitsaufnahme

Quantitation of mRNA by polymerase chain reaction :
nonradioactive PCR methods / Th. Köhler ... - Berlin ;
Heidelberg ; New York ; Barcelona ; Budapest ; Hong Kong ;
London ; Milan ; Paris ; Tokyo : Springer, 1995
  (Springer lab manual)
  ISBN-13: 978-3-642-79714-9
NE: Köhler, Thomas

ISBN-13: 978-3-642-79714-9   e-ISBN-13: 978-3-642-79712-5
DOI: 10.1007/978-3-642-79712-5

Production: PRODUserv Springer Produktions-Gesellschaft, Berlin
Typesetting: Fotosatz-Service Köhler OHG, Würzburg
Cover-layout: Struve & Partner, Heidelberg

SPIN 10493899   52/3020- 5 4 3 2 1 0   Printed on acid-free paper

Preface

In the last decade, no comparable method has revolutionized both research and diagnosis in clinical laboratories more than the polymerase chain reaction (PCR). This simple technique, which was developed by scientists of the Cetus Corporation in 1985, allows the in vitro enzymatic synthesis of millions of copies of a specific DNA segment even starting with an incredibly small number of target molecules. Originally developed for the identification and sequencing of novel genes and pathogens within the Human Genome Project (HUGO) this method has been applied more and more to medical disciplines, for example diagnosis and screening of genetic diseases and cancer, detection of minimal residual disease and drug resistance of tumors, the rapid detection of slow growing bacteria and difficult to cultivate viruses, and in HLA typing.

More recently and with increasing success, PCR and related techniques are used for the quantitation of viral genomes in body fluids and determination of the relative levels of abundance of a particular mRNA, changes in its abundance over time or after induction, and the actual number of mRNA molecules in tissue samples. The latter attempt is of increasing interest because PCR techniques have been shown to be thousands of times more sensitive than traditional attempts, e.g. Northern blot. This exquisite sensitivity gives us the ability to detect and quantify extremely rare mRNAs, or mRNAs in small numbers of cells or amounts of tissue with the additional advantage that only up to one day is needed.

This book is intended to provide a practical introduction into strategies using PCR-based techniques (e.g. competitive PCR, ELOSA-Enzyme Linked Oligonucleotide Sorbent Assay) to quantify mRNAs in biological samples. It was originally designed as a course manual for our traditional annual practical workshop for molecular biology, which last year took place with participants from 13 European countries and Israel and was supported by FEBS. We were encouraged by the participating students to publish these applied optimized protocols. In addition to the experimental parts, theory and applicability of each method are illustrated by some new figures and the advantages and limitations associated with them are compared.

It is emphasized that exclusively non-radioactive detection protocols, for instance HPLC, biotin and digoxigenin based detection protocols, are described in this manual, thus preventing time consuming licensing procedures and expensive discharge of radioactive waste.

This volume is an attempt to combine all the practical experiences and discussions with the course participants from the summary of the courses. Thus this laboratory guide should be helpful for both professionals and beginners.

The authors wish to emphasize that all listed chemicals, kits and technical supports proved themselves to be very worthwhile in the respective laboratories. Nevertheless, all the materials may be replaced by other products of comparable quality.

Finally we wish to thank Springer-Verlag and the sponsoring companies for their support of the book.

Leipzig, August 1995

Th. Köhler
D. Laßner
A.-K. Rost
B. Thamm
B. Pustowoit
H. Remke

Abbreviations

ADR	Adriamycine
AMV	Avian myeloblastosis virus
AP	Alkaline phosphatase
bp	base pair
BSA	Bovine serum albumin
cDNA	complementary DNA
CFTR	Cystic-fibrosis transmembrane-conductance regulator
CMV	Cytomegalovirus
cpm	counts per minute
cRNA	copy RNA (i.e. cDNA complementary RNA synthesized from synthetic DNA constructs)
CSPD	Disodium 3-(4-methoxyspirol{1,2-dioxetane-3,2′-(5′-chloro) tricyclo [3.3.1.1$^{3.7}$]decan}-4-yl)phenyl phosphate
dATP	2′-deoxyadenosine 5′-triphosphate
dCTP	2′-deoxycytidine 5′-triphosphate
ddNTPs	2′,3′-dideoxynucleotide triphosphate
DEAE	Diethylaminoethyl
DEIA	DNA Enzyme Immunoassay
DEPC	Diethylpyrocarbonate
dGTP	2′-deoxyguanosine 5′-triphosphate
DIG	Digoxigenin
DMFA	Dimethyl formamide
DMSO	Dimethyl sulfoxide
DNase	Deoxyribonuclease
dNTP	Deoxynucleoside triphosphates
DTT	Dithiothreitol
dTTP	2′-deoxythymidine 5′-triphosphate
dUTP	2′-deoxyuridine 5′-triphosphate
EDTA	Ethylenediamine tetraacetic acid
EGTA	Ethylene glycol-bis-(β-aminoethyl ether)-tetraacetic acid
ELISA	Enzyme-linked immunosorbent assay
ELOSA	Enzyme-linked oligonucleotide sorbent assay
GAPDH	Glyceraldehyde-3-phosphate dehydrogenase
HPLC	High performance liquid chromatography
kb	kilobase
K_D	Dissociation constant

LiDS	Lithium dodecyl sulphate
Lumigen PPD	4-methoxy-4-(phosphatphenyl)spiro(1,2-dioxetan-3,2'-adamantan)
MCS	Multiple cloning site in plasmids
MDR	Multidrug resistance
MOPS	Morpholinopropanesulfonic acid
MPC	Magnetic particle concentrator
mRNA	Messenger ribonucleic acid
MRP	Multidrug resistance-associated protein
MW	Molecular weight
NaN_3	Sodium azide
NBT	Nitro blue tetrazolium
nt	Nucleotide
O.D.	Optical density
p53	Tumor suppressor gene p53
PAA	Polyacryl amide
PAGE	Polyacryl amide gel electrophoresis
PBS	Phosphate buffered saline
PCR	Polymerase chain reaction
POD	Peroxidase (horseradish)
RARα	Retinoic acid receptor α
RARβ	Retinoic acid receptor β
RARγ	Retinoic acid receptor γ
rpm	Rounds per minute
rRNA	Ribosomal RNA
RT	Reverse transcriptase
SDS	Sodium dodecyl sulfate
SRY	Sex-determining region Y gene
SSC	Standard saline citrate
ssDNA	Single stranded DNA
TAE	Tris-acetate-EDTA
TBE	Tris-borate-EDTA
TE	Tris-EDTA
T_m	Melting point of DNA
TMB	Tetramethylbenzidine
tRNA	transfer RNA
Tth pol	DNA polymerase of Thermus thermophilus
U	Unit
UDG	Uracil-DNA glycosylase
UV	Ultra violett
VCR	Vincristine
X-Gal	5-bromo-4-chloro-3-indolyl-β–D-galactoside
X-phosphate	5-bromo-4-chloro-3-indolylphosphate

Trademarks

AmpliTaq™, GeneAmp™ 9600 thermal cycler, MicroAmp™ reaction tubes, GeneAmp® Thermostable rTth Reverse Transcriptase RNA PCR Kit (Perkin-Elmer Co., Norwalk, U.S.A.)

CleanGel™, ImageMaster™ Software, Sephaglas™ BandPrep Kit, MicroSpin™ Colums (Pharmacia LKB Biotechnology, Uppsala, Sweden)

EMBL® (European Molecular Biology Laboratory, Heidelberg, Germany)

GenBank® (National Institutes of Health, USA)

Hitachi HIBIO DNASIS™ DNA Sequence Analysis System (Hitachi Software Engineering Co. Ltd., Yokohama, Japan)

Medline® U.S. National Library of Medicine (Silver Platter International N.V.)

OLIGO® 5.0 Primer Analysis Software for Windows (National Biosciences, Inc.)

TaqStart™ monoclonal antibody (Clontech Laboratories Inc., Palo Alto, U.S.A.)

ScanPack® (Biometra, Göttingen, Germany)

Wizard™ PCR Preps DNA Purification System, PolyATtract® mRNA Isolation System, fmol™ (Promega Corporation, Madison, U.S.A.)

Tween®20 (ICI Americas Inc.,Wilmington, U.S.A.)

Lumigen™ PPD (Lumigen Inc., Detroit, U.S.A.)

Hyperfilm™-ECL (Amersham, Little Chalfont, U.K.)

CSPD™, SEQ-Light™, AVIDx™-AP and I-Block™ (Tropix Inc., Bedford, U.S.A.)

Polaroid® (Polaroid Corporation, Cambridge, MA, U.S.A)

Bacto®-Agar, Bacto® Yeast Extract (Difco, Detroit, U.S.A.)

RNAzol™ B, Trisolv™ (Biotecx Inc., Houston, U.S.A.)

OneShot™ competent cells, pCR II™ plasmid, TA Cloning® Kit (Invitrogen Corp., San Diego, U.S.A.)

TRIzol™ (Life Technologies Inc. Gibco BRL, Gaitherburg, U.S.A.)

Dynabeads®, Dynabeads® mRNA DIRECT™ Kit (Dynal A.S., Oslo, Norway)

Roti®Phenol (Roth, Karlsruhe, Germany)

QIAprep™ Spin Plasmid Kit (Qiagen GmbH, Hilden, Germany)

Sequenase™ (United States Biochemical, Cleveland, U.S.A.)

PreMix Long Ranger™ Gel (AT Biochem Inc., Malvern, U.S.A.)

Gen-Eti-K™ (Sorin Biomedica, Saluggia, Italy)

PCR is covered by US patents numbered 4 683 202, 4 683 195, and 4 965 188 issued to Cetus Corporation and owned by Hoffmann La-Roche.

Introduction

In many applications in biomedical research and clinical diagnosis it may be necessary to measure the concentration of a given target DNA or RNA. The need for accurate evaluations is important for clinical applications such as gene expression studies [1–6] e.g. the determination of "multidrug-resistance" gene expression where assessment of drug effectiveness is required following treatment with cytotoxic drugs or for detection of microbial infections [7, 8] e.g. measurement of viral RNA levels in patients with asymptomatic hepatitis C. Here the estimation of the exact copy number of a nucleic acid can be helpful in evaluating the physiological role of the investigated transcript or the success of therapy. Even though relative quantitation can be used to judge the efficacy of treatment, absolute quantitation is a better reflection of the patient's clinical state. However, small amounts of tissue often limit broad examination of gene expression for routine clinical investigation. Therefore, a methodology is needed that can provide sensitive and quantitative detection of absolute numbers of target molecules.

PCR, as a powerful and sensitive tool, can be employed to measure unknown concentrations of DNA (e.g. viral or bacterial DNA) or mRNA in a sample. But resulting from its exponential nature, some limitations of this method must be taken into consideration to prevent experimental inaccuracies in processing the method which could result in a potentiation of each failure [9, 10].

Amplification of mRNA molecules to study gene expression can be achieved by a method that combines two sequential enzymatic steps: the synthesis of DNA from the RNA template by reverse transcriptase (RT) and followed by PCR using of a heat stable DNA polymerase (Taq polymerase). The extremely high sensitivity of RT-PCR enables us to detect rare mRNAs, mRNAs in small numbers of cells or in small amounts of tissue as well as mRNAs expressed in mixed-cell populations. Here conventional techniques e.g. Northern blotting which requires at least 10^5–10^6 molecules per sample [11, 12], in situ hybridization [13] or solution hybridization (e.g. nuclease protection assay) [14] often fail to provide quantitative results. For instance the detection of interleukin 5 (IL-5) mRNA was shown to be impossible to achieve using conventional Northern blot analysis [15].

To obtain accurate quantitative results with this two-step technique is still problematic but possible with some adaptations that will be discussed in the following chapters. Nevertheless, the unlimited possibilities for application are leading to increasing efforts to use quantitative PCR-based strategies for diagnostic and research fields. This trend is partially reflected by the rise of relevant publications from only 3 in 1991 to 73 in 1994 (excluding December) as analyzed by the Medline Bibliographic Data Base for medicine.

To date some different quantitative RT-PCR protocols have been described to determine relative levels of diverse human mRNAs, e.g. for mdr-1 [1, 16–19], CFTR [18], cardiac muscle sodium channel [20], cytotoxic cell proteinases [21], dystrophin [2], various cytokins and cytokin receptors [3, 11, 15], several oncogens [12], renin [4] etc. These new approaches have also been tested to determine the activity of heterologous promoters by directly measuring reporter gene transcription products [23], to quantitate basal reporter gene expression in transient transfection assay [24] and as a very recent and forward-looking attempt to measure the transduction efficiency for gene therapy [25].

Based on our experiences in the detection and measurement of gene expression involved in the formation of the "Multidrug resistance" phenotype or genes responsible for degradation and rebuilding of the connective tissue in association with rheumatoid arthritis, we want to provide a collection of quantitative PCR based methods with this laboratory guide. The practically tested reliable protocols are discussed critically for their use and trouble shooting associated with them. We have tried to highlight the prospective trends of PCR related quantitation methodologies for clinical analysis.

References

1. Murphy LD, Herzog CE, Rudick JB, Fojo AT, Bates SE (1990) Use of polymerase chain reaction in the quantitation of mdr-1 gene expression. Biochemistry; 29:10351–10356
2. Chelly J, Montarras D, Pinset C, Berwald-Netter Y, Kaplan J-C, Kahn A (1990) Quantitative estimation of minor mRNAs by cDNA-polymerase chain reaction. Eur J Biochem; 187:691–698
3. Wang AM., Doyle MV, Mark DF (1989) Quantitation of mRNA by the polymerase chain reaction. Proc Natl Acad Sci USA; 86:9717–9721
4. Caroff N, Della Brunna R, Philippe J, Corvol P, Pinet F (1993) Regulation of human renin secretion and renin transcription by quantitative PCR in cultured chorionic cells: Synergistic effect of cyclic AMP and protein kinase C. Biochem. Biophys Res Comm; 193:1332–1338
5. Nagano M, Kelly PA (1994) Tissue distribution and regulation of rat prolactin receptor gene expression. J Biol Chem; 269:13337–13345
6. Yokoi H, Natsuyama S, Iwai M, Noda Y, Mori T, Mori KJ, Fujita K, Nakayama H, Fujita J (1993) Non-radioisotopic quantitative RT-PCR to detect changes in mRNA levels during early mouse embryo development. Biochem Biophys Res Comm; 195:769–775
7. Jalava T, Lehtovaara P, Kallio A, Rauki M, Söderlund H (1993) Quantification of hepatitis B virus DNA by competitive amplification and hybridization on microplates. BioTechniques; 15:134–139
8. Miller RH, Bukh J, Purcell RH (1993) Importance of the polymerase chain reaction in the study of hepatitis C virus infection. Int J Clin Lab Res; 23:139–145
9. Katz ED, DiCesare JL, Picozza E, Enderson MS (1993) General aspects of PCR quantitation. Amplifications; 10:7–8
10. Siebert PD (1993) Quantitative RT-PCR. Methods & Applications; Book 3, Clonetech Laboratories, Inc.
11. Bouaboula M, Legoux P, Pessegues B, Delpech B, Dumont X, Piechaczyk M, Casellas P, Shire D (1992) Standardization of mRNA titration using a polymerase chain reaction method involving co-amplification with a multispecific internal control. J Biol Chem; 267:21830–21838

12. Scheuermann RH, Bauer SR (1993) Polymerase chain reaction-based mRNA quantification using an internal standard: Analysis of oncogene expression. Methods Enzymol; 218: 446–473
13. Nonradioactive in situ hybridization. Application manual. Boehringer Mannheim GmbH 1992.
14. Farrell RE (1993) Quantitation of specific messenger RNAs by the S1 nuclease assay. In: Harcourt, Brace, Jovanovich (eds.). RNA methodologies. A laboratory guide for isolation and characterization. Academic Press, San Diego, New York, pp. 221–234
15. Guiffre A, Atkinson K, Kearney P (1993) A quantitative polymerase chain reaction assay for interleukin 5 messenger RNA. Anal Biochem; 212:50–57
16. Lyttelton MP, Hart S, Ganeshaguru K, Hoffbrand AV, Mehta AB (1994) Quantitation of multi-drug resistant MDR1 transcript in acute myeloid leukaemia by non-isotopic quantitative cDNA-polymerase chain reaction. Br J Haematol; 86:540–546
17. Lönn U, Lönn S, Nylen U, Stenkvist B (1992) Appearance and detection of multible copies of the mdr-1 gene in clinical samples of mammary carcinoma. Int J Cancer ; 51:682–686
18. Bremer S, Hoof T, Wilke M, Busche R, Scholte B, Riordan, JR, Mass G, Tümmler B (1992) Quantitative expression patterns of multidrug-resistance P-glycoprotein (MDR1) and differentially spliced cystic-fibrosis transmembrane-conductance regulator mRNA transcripts in human epithelia. Eur J Biochem; 206:137–149
19. Futscher BW, Blake LL, Gerlach JH, Grogan TM, Dalton WS (1993) Quantitative polymerase chain reaction analysis of mdr1 mRNA in multiple myeloma cell lines and clinical specimens. Anal Biochem; 213:414–421
20. Zhou J, Hoffman EP (1994) Pathophysiology of sodium channelopathies. J Biol Chem; 269: 18563–18571
21. Prendergast JA, Helgason CD, Bleackley RC (1992) Quantitative polymerase chain reaction analysis of cytotoxic cell proteinase gene transcripts in T cells. J Biol Chem; 267:5090–5095
22. Caroff N, Della Brunna R, Philippe J, Corvol P, Pinet F (1993) Regulation of human renin secretion and renin transcription by quantitative PCR in cultured chorionic cells: Synergistic effect of cyclic AMP and protein kinase C. Biochem Biophys Res Comm; 193: 1332–1338
23. Kovacs DM, Kaplan BB (1992) Discordant estimates of heterologous promoter activity as determined by reporter gene mRNA levels and enzyme activity. Biochem Biophys Res Comm; 189:912–918
24. Morales MJ, Gottlieb DI (1993) A polymerase chain reaction-based method for detection and quantification of reporter gene expression in transient transfection assays. Anal Biochem; 210:188–194
25. Rettinger SD, Kennedy SC, Wu X, Saylors RL, Hafenrichter DG, Flye MW, Ponder KP (1994) Liver-directed gene therapy: Quantitative evaluation of promoter elements by using in vitro retroviral transduction. Proc Natl Acad Sci; 91:1460–1464

Contents

Chapter 2.2
Synthesis of cDNA   65

D. LASSNER

Chapter 2.3
Qualitative RT-PCR: Amplification of Synthesized mdr-1 cDNA   71

TH. KÖHLER

Chapter 2.4
Single-Tube RT-PCR   81

D. LASSNER

Chapter 2.5
Nonradioactive Determination of PCR Products by Using
a DIG-Labeled DNA Probe (Dot Blot)   85

TH. KÖHLER

Chapter 2.6
Nonradioactive Northern Blot Hybridization with DIG-Labeled DNA Probes . . . 93

A.-K. ROST

Part 3
Semiquantitative and Quantitative Protocols for Measurement
of Nucleic Acids by PCR 115

Chapter 3.1
Quantitation of mRNA by the ELOSA Technique Using External Standards 117

D. LASSNER

Chapter 3.2
Semiquantitative Detection of Viral DNA, e. g. for CMV,
by Using the DNA Enzyme Immunoassay (DEIA) 125

B. Pustowoit

Chapter 3.3
HPLC – Analysis of Nucleic Acids 135

H. Remke and Th. Köhler

Chapter 3.4
Quantitation of Absolute Numbers of mRNA Copies in a cDNA Sample
by Competitive PCR 143

Th. Köhler

Part 1

Theoretical and Methodical Prerequisites for Using PCR to Quantitate Nucleic Acids

Chapter 1.1
General Aspects and Chances of Nucleic Acid Quantitation by PCR

TH. KÖHLER

1.1.1 Theory of Template Amplification by PCR

Quantitative PCR analysis compared with qualitative PCR reaction is more complicated because of two features inherent in in vitro amplification. First, during the exponential phase of reaction minute differences in a number of variables (e.g. concentration of reaction components, tube-to-tube variation) can greatly influence the reaction rates, with a substantial effect on PCR product yield. Second, as a consequence of reaction component consumption and generation of inhibitors (e.g. pyrophosphates, final PCR products) the reaction enters a plateau phase where the reaction rate declines dramatically.

1.1.1.1 Mathematical Description of the PCR Reaction

An optimized amplification protocol is fundamental for quantitation of nucleic acids by PCR. Optimization of a PCR reaction is not problematic in the majority of cases and should therefore not be subject of this book. We therefore recommend reading the ongoing literature [1, 2, 44, 45].

By definition, the PCR process is a chain reaction. The products from one cycle of amplification serve as template for the next leading to an exponential and non-linear increase of the product. Under theoretical conditions, the amount of product doubles during each cycle of PCR according to Eq. (1).

(1) $\qquad N = N_o * 2^n$

$\qquad$ N $\ =$ number of amplified molecules
$\qquad$ $N_o =$ initial number of molecules
$\qquad$ n $\ =$ number of amplification cycles

A crucial factor for the reliability of quantitative PCR is the amplification efficiency. That is the fraction of the template which is replicated during each reaction cycle. Experimentally, the efficiency of amplification is less than ideal (Figure 1.1-1) and may vary in the exponential phase of reaction from 0.78 – 0.97 for different genes (3 – 6). Very small differences in the amplifica-

log peak area [mV*s]

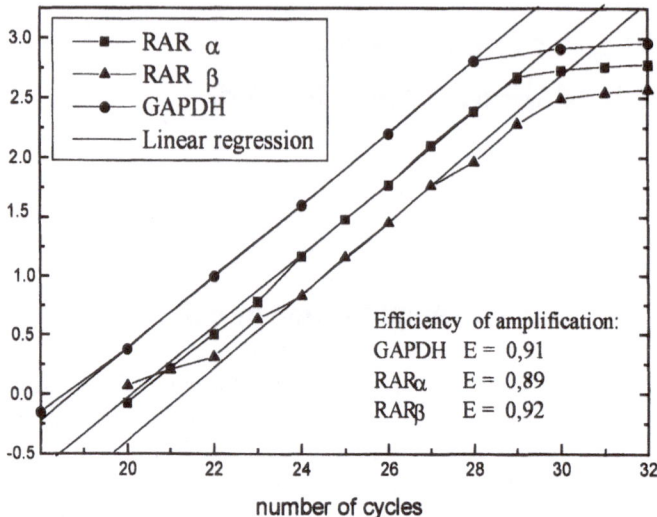

number of cycles

Fig. 1.1-1. Ascertainment of efficiency for amplification of 3 different genes by HPLC analysis of PCR fragments, the plateau phase is reached between cycle 28 and 30, respectively. Values for the efficiency E of the amplification may be calculated from the slope of the curves following linear regression analysis

tion efficiency, therefore, can have profound changes on the abundance of the final product. The efficiency (E) of the reaction is influenced by several factors:

- concentrations of DNA polymerase, dNTPs, $MgCl_2$
- PCR product strand reannealing during later amplification cycles leading to a diminished efficiency of the reaction
- secondary structure and G/C content of target sequence affect the melting point of the target DNA, interfere with primer binding, and reduce processibility of Taq polymerase
- sequence and composition of chosen primers strongly influence the primer annealing kinetics
- insufficient optimization of the PCR protocol, occurrence of side products
- length of sequence being amplified: synthesized fragments should not be longer than 1 000 bp, because E is roughly inversely proportional to the fragment length separating the two primers on their analyzed template [3]
- presence of inhibitors of reverse transcriptase or Taq polymerase in the RNA preparations
- tube to tube effects from unknown reasons

The process of product accumulation is therefore better described by Eq. (2).

(2) $N = N_o * (1 + E)^n$

where: E = amplification efficiency

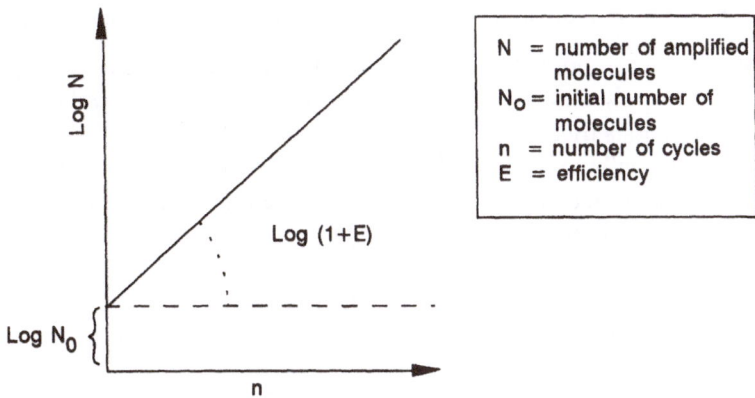

Fig. 1.1-2. Kinetic estimation of the initial number of molecules (N_0) in a sample and efficiency (E) of PCR amplification reaction

In practice, equation 2 is utilized in its logarithmic form:

(2.1) $\text{Log } N = \text{Log } N_0 + n * \text{Log } (1 + E)$

By plotting the logarithmic values of the PCR product yield (Log N) against the number of amplification cycles (n) a linear curve can be generated with the intercept equal to the logarithmic values of the target (Log N_0) and the slope equal to Log $(1 + E)$ (Figure 1.1-2). Equation (2.1) is only valid if PCR is performed in the exponential phase of the reaction and is typically used to assess the cycle-to-cycle efficiency of amplification [7].

Equation (2) can be also rearranged as:

(2.2) $N_0 = N/(1+E)^n$

When E is deduced from the slope of the semi-log plot (Eq. (2.1)) and the efficiency remains constant (exponential phase of reaction) the value of N_0 can be calculated from Eq. (2.2) [3, 8].

1.1.1.2 The Plateau Phase of Reaction

Experimentally the kinetics of product accumulation are far from the theoretical case as described by Eq. (2) and shown in Figure 1.1-2 because amplification is not exponential over the whole cycle range. The amount of product initially increases exponentially with efficiencies close to 1. During later amplification cycles, the rate of production slows. This effect is referred to as the "plateau effect" or "saturation" [1, 9, 10].

Factors that may be responsible for this observation are as follows:

- One or more of the components necessary for the reaction become limiting, especially the DNA polymerase, which is the most expensive component of the reaction mixture and is therefore employed with only a slight excess.
- The product or partially extended primers accumulate to a concentration at which their reassociation competes with primer annealing and extension [11].
- Pyrophosphates which are known as inhibitors of polymerase activity accumulate [1].
- The thermal inactivation of polymerase during later cycles (e. g. the half live of AmpliTaq distributed by the Perkin-Elmer Co., is approximately 35 min at 95 °C) [9].
- The denaturation efficiency is reduced especially in the later cycles [9].

It should be mentioned here that simultaneous amplification of two or more different targets using the same primer pair (e.g. as introduced by competitive PCR, see 1.1.2.2) will result in reaching the plateau phase more rapidly at much lower cycle numbers.

1.1.2 Experimental Approaches to Using PCR for Quantitation of mRNA

The first steps in the analysis of mRNA are isolation (see chapter 2.1) and the reverse transcription into cDNA (see chapter 2.2) serving as template for PCR. Following amplification, the goal of quantitative PCR is to draw a conclusion from the amount of final PCR product to the initial number of target molecules or the relative starting levels of target molecules among several samples. Various approaches have been developed in the last few years by inclusion of external and internal standards.

1.1.2.1 Quantitative PCR Using External Standards

Quantitation of nucleic acids by PCR assays using external standards is generally possible, however, it is inconvenient for two restrictive reasons: tube-to-tube variations regarding amplification efficiency must be minimal and all data must be obtained before the reaction reaches the plateau. Thus, complex and time-consuming operations are required before the actual quantitation experiments can be started.

If the two conditions stated above are in effect, generally speaking two forms of experimental analysis, namely titration and kinetics, can be used to estimate the relative initial amounts of target mRNA or cDNA in different samples.

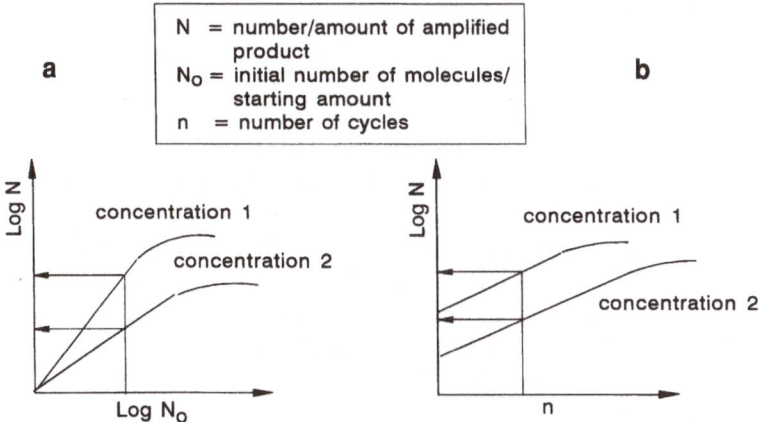

Fig. 1.1-3. Methods for the experimental determination of relative differences in the initial amount of target molecules of two samples (according to Siebert [1]). **a** Titration analysis, **b** Kinetic analysis

Titration Analysis

This type of PCR analysis is performed using several dilutions of analyzed cDNA of unknown concentration. In parallel a set of dilutions of a cDNA with known concentration or a reference sample is prepared. Following amplification in separate tubes and measuring the product the Log of the amount of synthesized DNA (N) is graphed as a function of the initial amount of the reference sample (N_0). Because in the exponential phase of the reaction the amount of amplified DNA is a constant proportion of the total starting material (N_0) for each of the various dilutions of a sample, the relative difference in N_0 between two samples is proportional to the difference between the slopes of the curves (Figure 1.1-3, Panel a).

Thus, for quantitation, a value of N_0 is chosen on the X axis and the corresponding value of Log N are extrapolated for both curves. The difference between the two values is equivalent to the relative differences in N_0 for the two samples (Figure 1.1-3, Panel a).

Kinetic Analysis

In contrast to the titration analysis the kinetic analysis is the more commonly used method. Values of N are determined for a number of consecutive amplification cycles (n) for different samples (Figure 1.1-3, Panel b). A value of n is chosen at a point where the two curves are parallel, suggesting equal efficiency of the PCR reaction. The values of Log N are extrapolated as described above. At the chosen point the difference between the two values of Log N is directly proportional to the difference of Log N_0 between the two samples. However, in contrast to the titration method, this type of analysis can only be used to estimate differences in the initial number, not the initial number of target molecules itself.

1.1.2.2 Quantitative PCR Using Internal Standards

To use internal standards for quantitative PCR, two different methods may be used. The first is the use of endogenous housekeeping gene transcripts which are contained in every RNA preparation derived e.g. from eucaryotic cells. These internal control-mRNAs may be used under the assumption that the coding genes are transcribed constantly and independently from the extracellular environment and that their transcripts are reverse transcribed with the same efficiency as the product of interest. In the second method, an exogenous native or synthetic mRNA or DNA standard of known concentration is added to the determined target (Figure 1.1-4). Both approaches share the property that target and standard are amplified simultaneously in the same reaction tube.

Researchers most commonly use such internal standards to control tube-to-tube effects on amplification efficiency and to determine absolute amounts of mRNAs.

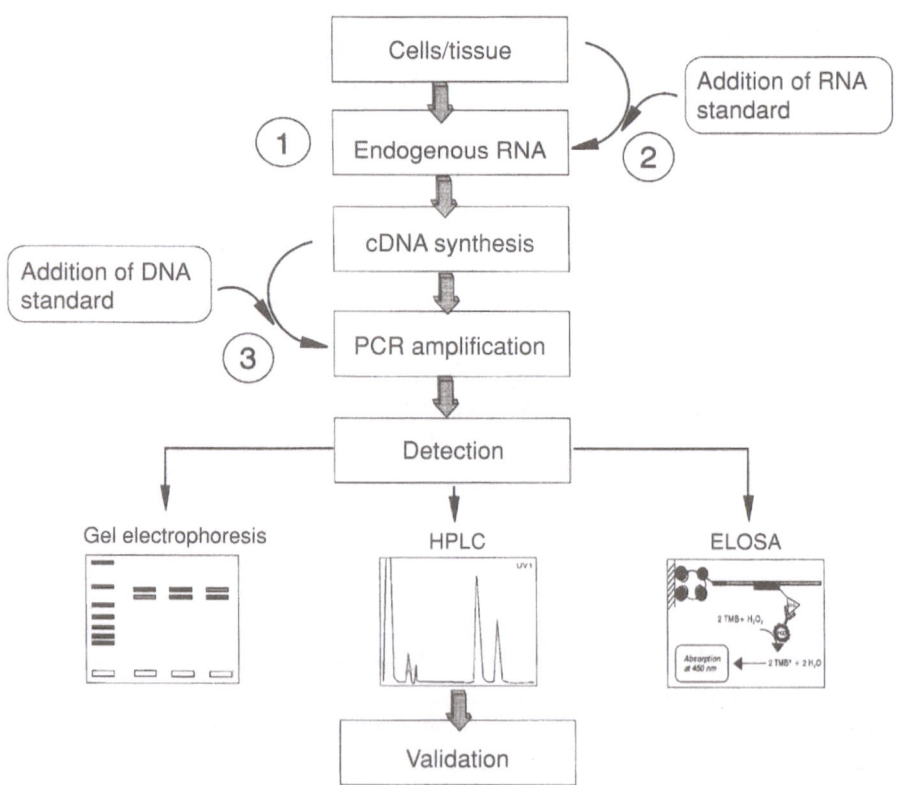

Fig. 1.1-4. Different approaches to use internal PCR standards. (*1*) Housekeeping genes are used as endogenous control, (*2*) cRNA standard added prior to RNA preparation or reverse transcription, (*3*) DNA standard added prior to PCR reaction

Endogenous mRNA as Internal Control

One of the first attempts to use PCR to quantitate mRNAs of interest was the evaluation of relative amounts of target mRNA in a sample by coamplifying a structurally unrelated endogenous mRNA, e.g. for β-actin [12–15], β₂ micro-globulin [16, 17], aldolase A [3, 18], elongation factor EF-1α [19] or GAPDH [10, 20–22]. But this approach is problematic when the control gene is affected by experimental or physiological conditions, e.g. as described for the β-actin gene in human fibroblasts [10]. Only the genes for histone H 3.3 or ribosomal protein L19 (rpL19) have shown to be cell-cycle independent and constitutively expressed in all tissues [23, 24]. Either in two separate PCR reactions or amplified in the same reaction, a minimization of tube-to-tube variations may be achieved. However, when amplifying target and reference mRNA in the same tube or not, both templates may be amplified with different efficiencies and the point when the reaction reaches the plateau must be known to stop the amplification during its exponential phase. These are the main prerequisites for obtaining reliable quantitative results. The data can be obtained either by titration or kinetic analysis as described previously.

If endogenous target and internal standard are simultaneously amplified the amounts of both products generated following n cycles would be according to Eq. (2):

$$N_t = N_{0t} (1 + E_t)^n$$
$$\text{and } N_s = N_{0s} (1 + E_s)^n$$

Both terms combined give Eq. (3).

(3) $N_t/N_s = N_{0t} (1 + E_t)^n / N_{0s} (1 + E_s)^n$ or rearranged
       $\text{Log } N_t/N_s = (N_{0t}/N_{0s}) + n * \text{Log } (1 + E_t)/(1 + E_s)$

where: N_{0t} = initial number of target molecules
          N_{0s} = initial number of standard molecules
          N_t = amount (number) of amplified target molecules
          N_s = amount (number) of endogenous standard molecules
          E_s = efficiency of standard amplification
          E_t = efficiency of target amplification
          n = number of cycles

The relative initial amounts of target mRNA at a known concentration of endogenous standard can now be determined.

From Eq. (3) it is seen that even in the case where E_s is not equal to E_t endogenous mRNAs can be used to compare relative target levels. If the amplification efficiencies of both target and standard are identical, equation 3 can be simplified to Eq. (4):

(4) $N_{0t}/N_{0s} = N_t/N_s = A_t/A_s$

where: A_s = amount of amplified standard (cpm or OD_{260} units)
            A_t = amount of amplified target (cpm or OD_{260} units)

Attempts to use PCR, as described above, are time consuming because they require the estimation of the amplification efficiencies for each run that may vary from sample to sample even when the same transcript was amplified under identical conditions [1].

In practice these efforts are simplified in that it is not the ratios of the initial absolute numbers of target and standard that are estimated but the measured amounts of each PCR products in different samples are compared. For this reason this technique is often termed "semiquantitative" or "comparative RT-PCR" (cRT-PCR) [23] because it does not result in an absolute value for the mRNA of interest.

Another problem is that simultaneously amplified mRNAs may be initially present at widely different levels resulting in competition for limiting PCR substrates. The effect of such competition is a loss of exponential amplification and hence an attenuation of the final product levels [16, 23]. A clever solution to this problem is simply waiting before adding the primers for the endogenous standard at later amplification cycles (primer dropping method) and/or by selectively limiting the number of PCR cycles [10, 13] or modification of the concentration of each set of primers according to the relative abundance and size of the corresponding PCR product of each mRNA amplified [20].

But because of the associated problems, such experiments are somewhat imprecise and increasingly inconvenient in practice. Therefore, attempts have been made to develop quantitative procedures involving reactions driven to the plateau phase resulting in higher product accumulation combined with higher sensitivity and advanced reproducibility.

Competitive PCR Using Exogenous Added RNA/DNA as Internal Control

Competitive PCR is a quantitative adaptation of the PCR method in which a known number of copies of an exogenously synthesized cRNA [4, 5, 14, 17, 19, 21, 22, 24–28] or DNA [15, 29–32, 43, 46] are introduced together with the sample into the PCR reaction mixture. The basis of competitive PCR is the coamplification of a nucleic acid of interest and the internal standard DNA. This is possible if target and standard show the same DNA sequence or at least the same primer binding sites. Both targets therefore compete for the common primers and reagents in the same reaction tube. In competitive PCR, a dilution series is used either from the analyzed target or the competitor fragment and an identical amount of the other component is added to each of the reaction tubes. Both the amplified standard and product derived from the cDNA of interest are distinguished by either size, restriction endonuclease cleavage [22], or specific hybridization.

If the supposition is justified that the amplification efficiencies are similar, a relatively easy and accurate quantitation under plateau conditions is possible.

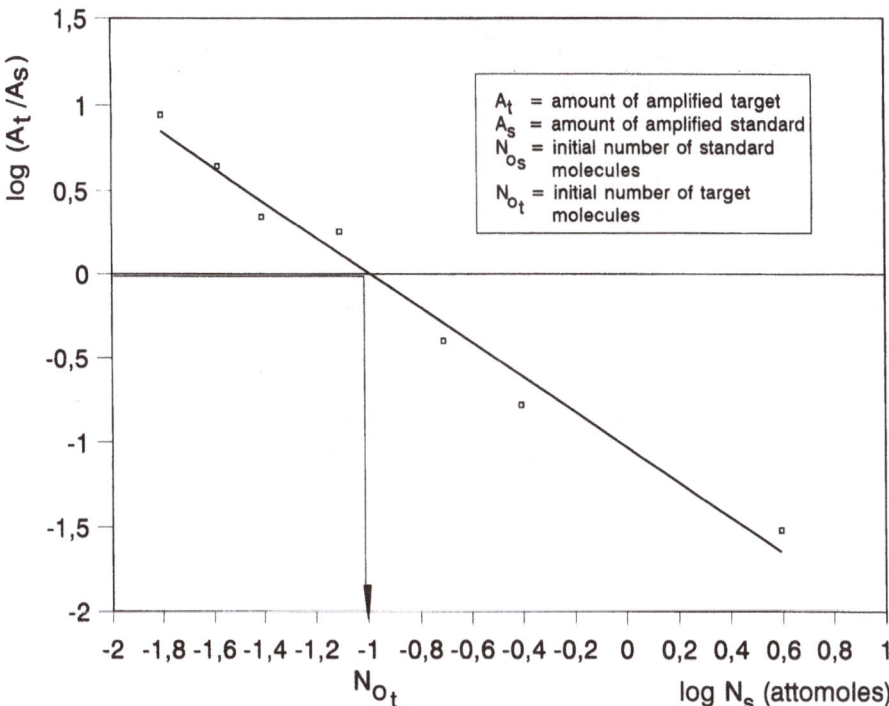

Fig. 1.1-5. Results of a competitive PCR experiment under plateau conditions. The Log of the ratio of amplified target to competitor product is graphed as a function of the Log of a known amount of competitor added to the PCR reaction. Note that when the molar ratio of target and competitor is equal to 1, the Log of ratio is equal to 0

The quantitation may be achieved by simply comparing the relative amounts of the two products A_t and A_s (Eq. (4), Figure 1.1-5).

But, as shown by Pannetier and coworkers [15] and according to our own observations, the internal standards must be very similar to the target DNA which is to be assayed in order to perform quantitation under saturating PCR conditions. Differences in length of only 4 bp out of a few hundred between standard and endogenous target may be sufficient to make quantitation at the plateau phase unreliable [15].

Because RNA extraction and reverse transcription often have rather variable yields [15, 21]), an additional adjustment (normalization) of the results is necessary. Since even the inclusion of an internal RNA standard at the reverse transcriptase reaction level would not check for any variations in RNA extraction, an assay analogous to one used for endogenous controls to measure housekeeping gene transcripts is absolutely essential in order to normalize the results [10, 15, 17, 19, 22, 46]. Housekeeping genes are assumed to be equally transcribed in various samples and their transcripts are reverse transcribed with comparable efficiencies. However, measurements of the control gene must also be performed in the exponential range. Since most of the control genes used are expressed at higher levels than most mRNAs under study, and therefore the reaction plateau is reached more quickly, non-comparative conditions may occur as described in section 1.1.2.2. But co-quantitation of a housekeeping transcript should be used to correct for RNA loading of the PCR reaction. This feature is very useful when amounts of total or poly A+ RNA cannot be quantified by UV spectrophotometry prior to RT-PCR [19].

Another interesting attempt to validate quantitative PCR results is the introduction of genomic DNA with known gene loci ratios. Zhou et al. [14] employed this method to measure sodium channel mRNA quantities in a sample which were corrected by amplified genomic DNA where the loci ratio of actin: adult: fetal sodium channel was known to be 1:1:1. The correction factor obtained was similar to that obtained using known amounts of cloned sodium channel gene plasmid DNA [14].

Although one of the best techniques now available a main disadvantage of competitive and most other quantitative PCR techniques exist: numerous PCR reactions are required to measure the level of a single target sequence within a sample. To overcome this ineffectiveness novel quantitation strategies are employed, e.g. quantitative multiplex fluorescent polymerase chain reaction (QMF-PCR) [14]. QMF-PCR was described as allowing the measurement of multiple mRNA species simultaneously in a single PCR reaction using primers labeled with fluorescent dyes in multiplex reactions (i.e. different sets of primers were simultaneously used to amplify a variety of different templates in the same tube) and analyzed on automated sequenators [14]. Two similar multiplex PCR based approaches were described by Wong et al. [10] and Dostal et al. [19] using a competitive multiplex PCR assay. The very recently described competitive and differential RT-PCR assay (CD-RT-PCR) [46] is a combination of multiplex PCR and normalization procedure.

However, single tube reactions containing more than one set of primers are limited by different primer kinetics and the fact that the exponential phase of the reaction is likely to occur after a different number of cycles [17, 26].

1.1.3 Sensitivity and Reproducibility of Quantitative RT-PCR Assays

To check the sensitivity of the procedure, reconstitution experiments are needed i.e. dilutions of a cloned target DNA were mixed with a constant amount of unrelated DNA and coamplified with known concentrations of an internal standard. 10 copies of e.g. HIV-DNA were readily assayed [15]. That means, rare mRNAs of about 0.003–0.1 copies per cell may be quantitated without any problems [6, 17, 26].

In spite of the method-associated problems, quantitative RT-PCR assays were shown to yield an accurate experimental precision of about 25% as assayed for a PCR assay run to saturation [15] and intraassay coefficients of about 2–7% [4, 13, 26]. Values obtained from both radioisotopic and non-radioisotopic quantitative RT-PCR assays were found to be consistent with values obtained from Northern blot analysis [4, 33].

Reverse transcription here is an important source of variability which may range from 5–90 % [15, 19, 31]. In most of the cases MMLV reverse transcriptase was used for reverse transcription because AMV RT was described to be less efficient in synthesis of cDNA [20]. The efficiency may be estimated e.g. by measurement of α^{32}P-dCTP incorporation into the newly synthesized cDNA [15]. The extraction of total RNA rather than polyA+ RNA was reported to decrease the risk of losing weakly expressed mRNAs. Dukas et al. [20] observed that oligo (dT) priming did not produce background PCR bands which appear by random priming whereas specific priming with the down-stream PCR primer is more efficient than oligo (dT) priming [26]. The occurrence of unspecific PCR product formation clearly affect the efficiency of PCR product accumulation.

1.1.4 Methods for Detection and Quantitation of PCR Products

Unlike the widespread use of radioactive labeled nucleotides or primers [13, 16, 23, 26, 31, 34] the PCR reaction and detection of products should be performed nonradioactively. The exact determination of molar concentrations of a product which accumulates in consecutive cycles can be performed by HPLC (chapter 3.3), ELOSA (chapter 3.1) or densitometric scanning of ethidium bromide stained gels following electrophoresis [19, 20]. The estimation of PCR product amounts from at least 3 different cycles in the exponential phase of reaction makes it possible to calculate the efficiency of amplification and initial amount of the target molecules.

1.1.5 Avoidance of PCR Contamination

The extremely high sensitivity of PCR makes this method susceptible to reagent and sample contamination or product "carryover". For this reason a number of routine precautions should be taken when performing any type of PCR to eliminate unwanted transfer of DNA to the reaction or to selectively destroy contaminating DNA. Often it will be sufficient to prevent carryover by following the general rules for "good laboratory practice":

- If possible set up the experiments in a laminar flow hood dedicated solely to PCR use.
- DNA amplification and PCR product storage, electrophoresis, purification or plasmid preparation should be situated in separate laboratories.
- Generally add the internal standards in a different room from the PCR reaction compartment.
- Separate pipettes, consumer materials, and reagents should be used for PCR, all materials should be autoclaved.
- Disposable gloves should be worn at all times and changed frequently.

Additional techniques should be employed to eliminate nevertheless occurring contamination. These efforts include subsequent DNase I treatment to eliminate trace DNA contaminants derived from plasmid DNA [35] or DNA crosslinking by exposing the contents of the PCR reaction mix to short-wave UV radiation before adding the sample template. Such an attack should nick and crosslink any contaminating DNA rendering them sufficiently unamplifiable [36].

The best results to prevent carryover we have achieved were by general substitution of TTP nucleotides with dUTP in conjunction with sample treatment with uracil-DNA glycosylase (UDG). All PCR products and the internal standard synthesized and used in our laboratory contain dUTP instead of dTTP. UDG added to later PCR reactions ensures deglycosylation of any contaminating DNA originating from previous PCRs, thus inhibiting its amplification [28, 37]. The extreme specificity of UDG for dU-containing DNA leaves the natural target unaffected. PCR products containing dU can be as well as natural DNA cleaved by restriction endonucleases [37], can be cloned efficiently using a UDG-deficient strain of *E. coli* [37] and can be amplified with the same efficiency compared to natural DNA.

References

(see end of chapter 1.2.)

Chapter 1.2
Design of Suitable Primers and Competitor Fragments for Quantitative PCR

Th. Köhler

1.2.1 Primer Selection

The first step to design specific primers is to get some sequence information from the gene or the mRNA of interest. All known sequences are available from special databases (e.g. EMBL, GenBank) which are commercially distributed by CD-ROM or via data networks, respectively. Using special hardware and software (e.g. Hitachi HIBIO DNASIS DNA Sequence Analysis System, vers. 7.0 or higher) partial or complete sequences can be picked out. Following selection of the sequence part of interest the primer search can be started.

Unfortunately, there is no general set of rules that will ensure the synthesis of an effective primer pair so that the selection is somewhat empirical. But the majority of primers can be made to work if the following guidelines are taken into account:

- Random base distribution over the total length of the primer typical primer length: 18–30 nucleotides
- G/C content between 50% and 60%, lower percentage of G/C is possible but cause lowered T_m
- T_m between 55°C and 80°C, balanced for a given primer pair
- Avoid sequences with significant secondary structure (e.g. loops, hairpins).
- Avoid complementary bases at the 3′ ends of primer pairs as this promotes the formation of primer dimers which compete with the target for Taq-polymerase and nucleotides.
- Test the specificity of the designed primer pair by running the selected sequence against the complete database.

Automated primer search software (e.g. OLIGO 5.0 Primer Analysis Software) in general follows these guidelines.

Primers designed for mRNA quantitation should be situated on two different exons and span at least one intron to allow unambiguous discrimination between cDNA and unwelcome genomic amplification products [25]. If genomic DNA is used as an internal control (see section 1.1.2.2) PCR primers have to be selected which may amplify both DNA and RNA (i.e. situated within a single exon) [14]. Another way to generate primers specific for mRNA detection is the situation of one of the primers directly on the fusion site of two neighboring exons which allows exclusive cDNA amplification.

To minimize the effect of possible variations in reverse transcription using oligo (dT) Zhou et al. [14] recommend choosing primers near the 3′ end of each coding sequence to be determined. Primers designed for quantitative PCR should generate PCR products in a window as narrow as possible to minimize preferential amplification of target sequences due to template size differences.

1.2.2 Design and Construction of Synthetic Internal PCR Standards

1.2.2.1 Synthetic Genes Serving as Multifunctional Standards (Multistandards)

As described by several groups [5, 21, 38] and for the first time by Wang et al. [4], synthetic genes were constructed facilitating the quantitation of different cellular transcripts. The constructs used as "multistandards" consisted of 5′ primer sequences of several target genes connected in sequence followed by the complementary sequences of the 3′ primers. These products were generated by sequential PCR reactions with overlapping primers and cloned using special vectors allowing insertion of the multiple primer regions between flanking RNA polymerase promoters and a downstream polyadenylated sequence. From the cloned and purified vectors the in vitro synthesis of cRNA serving as internal standard was performed (see section 1.1.2.2). Competitor fragments that share the same primer template sequence but contain a completely different inter-vening sequence were designated as heterologous competitors. As shown in Figure 1.2-1 and Table 1.2-1 a very similar construct (pMS1) showing the same features as described was designed in our laboratory. Linkers containing unique restriction enzyme recognition sites were placed after the sets of 5′ and 3′ primers to allow insertion of any additional primers needed. However, in contrast, the 3′ primer sequences were arranged in such an order that allows the synthesis of standard products of sizes comparable to the endogenous products thus yielding comparable amplification efficiencies. The insertion of a lacZ promoter element in the construct allows simple discrimination between the PCR products derived from standard and endogenous target by hybridization with a specific DNA probe or usage of lacZ promoter binding proteins e. g. lacI-β-galactosidase fusion protein as described by Lundeberg et al. [39].

Before synthesis, a newly designed standard nucleotide sequence should be analyzed for the presence of internal repeats and restriction sites because it was anticipated that internal repeats could cause problems during subsequent PCR amplification and quantitation experiments using internal standard RNA [21]. After synthesis and cloning a sequence analysis is necessary to find out point mutations generated during Taq polymerase synthesis which fidelity is far from optimal [11]. Longer overlapping oligonucleotides can be used to reduce the number of steps for the synthesis of the insert leading to a reduced number of errors [21]. An alternative way is the usage of Pfu DNA polymerase (Gibco) which in contrast does not possess terminal transferase activity (i.e. intro-duction of unwanted nucleotides, usually A, onto the 3′ ends of the newly

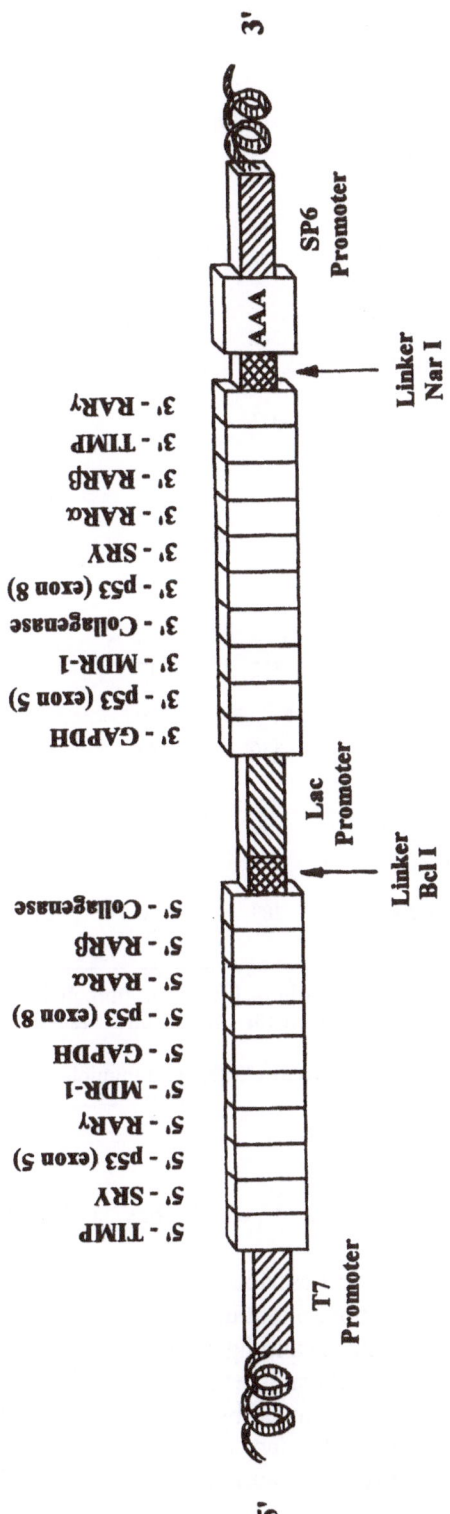

Fig. 1.2-1. Structure and composition of the pMS1 synthetic oligonucleotide

Table 1.2-1. Oligonucleotides used for quantitation of 9 different gene transcripts by competitive PCR

Gene	(+)-primer	(-)-primer	product size (bp)	standard size (bp)	Ref.
TIMP	5′ CAACCAGACC ACCTTATACC 3′	5′ AGTTCCACTC CGGGCAGGAT TC 3′	500	392	d)
SRY	5′ GAATATTCCC GCTCTCCGG 3′	5′ ACAACCTGTT GTCCAGTTGC 3′	422	311	11
GAPDH	5′ CGTCTTCACC ACCATGGAGA 3′	5′ CGGCCATCAC GCCACAGTTT 3′	300	139	a)
RARα	5′ GACCGGCTCT TGAGACATCC 3′	5′ GGTGCTGGAG AGTGGTCAGA 3′	371	212	b)
RARβ	5′ GGAACGCATT CGGAAGGCTT 3′	5′ GGAAGACGGA CTCGCAGTGT 3′	383	212	b)
RARγ	5′ CCAGTATGTA GAAGCCAGTC 3′	5′ GTAGCCAGAG GACTTGTCAT T	480	354	40
p53 (exon 5)	5′ TTCCTCTTCC TGCAGTACTC 3′	5′ GCAAATTTCC TTCCACTCGG 3′	325	217	41
p53 (exon 8)	5′ CCTATCCTGA GTAGTGGTAA 3′	5′ CCAAGACTTA GTACCTGAAG 3′	330	194	41
Collagenase	5′ TTGCACTGAG AAAGAAGA 3′	5′ TTGCCTCCCA TCATTCTT 3′	179	116	c)
mdr-1	5′ AGATCAACTC GTAGGAGTGT 3′	5′ GGGCTAGAAA CAATAGTG 3′	289	196	d)

a) Sequences given by Dr. H. Garn, Philipps-University of Marburg, Institute of Immunology.
b) Sequences given by Dr. S. Grams, GSF Forschungszentrum für Umwelt und Gesundheit GmbH, Neuherberg.
c) Sequences given by Dr. U. Sack, University of Leipzig, Institute of Clinical Immunology and Blood Transfusion.
d) Sequences designed in our laboratory.

synthesized double-stranded DNA molecule) and which has a higher polymerization fidelity [23].

Another way to construct synthetic genes was described by Bouaboula et al. [5]. The constructs were generated from the primer oligonucleotides by sequential phosphorylation with T4 polynucleotide kinase and ligation with T4 ligase. Only the last assembly step was performed by PCR using primers corresponding to the two extremities of the assembly and being flanked by EcoRI and HindIII restriction sites necessary for further cloning. But just as described for the former method, insertion errors cannot be prevented and have to be corrected e.g. by site-directed mutagenesis [5, 21].

1.2.2.2 Construction of Competitors by Site-Directed Mutagenesis

Methods Using PCR and Subsequent Cloning Strategies

Many PCR based methods for site-directed mutagenesis have been reported as being quicker and more convenient than non-PCR based mutagenesis [14, 19, 22–24]. Synthetic standards that produce a PCR product of a different size may be constructed by simple insertion of any one DNA sequence into compatible cloning sites located between the primer binding sites [24, 27]. Site-directed mutagenesis may be also performed by using mismatched primers to introduce a new restriction site into a PCR fragment [22]. Cloned and in vitro transcribed, the coamplified standard can be discriminated from the wild-type fragment by subjecting aliquots of the PCR products to restriction digestion.

However, quantitation experiments must be always performed during the exponential phase. Reassociation of DNA molecules not serving as template (e.g. in the plateau phase) could lead to heterodimers that are resistant to cleavage thus affecting the results [22]. An alternative protocol described the generation of deletions in cloned cDNA fragments by using inverse PCR. Deletions in cDNA were generated by amplification with primers which had a corresponding gap between their 5' ends [19].

A reliable method which yields up competitor fragments with only slight standard to target sequence differences was described by Zhou et al. [14]. Here PCR products or synthetic oligonucleotides were cloned into suitable vector plasmids. The plasmids were digested with appropriate restriction enzymes, the overhangs were filled in by Klenow enzyme or shortened by exonuclease activity of T4 DNA polymerase yielding synthetic fragments which differ slightly in size from the native fragment. Followed by in vitro blunt-end ligation (see chapter 1.3) the constructs were subcloned into a vector containing the T3/T7 polymerase promoter. In vitro transcription generated cRNAs with high sequence homology to the target sequence. Another efficient PCR based mutagenesis method which requires only one new primer for each desired mutation was described by Chen and Przybyla [23]. Starting with a vector that contains the gene of interest and the target mutation site both flanked by two known primer sequences a mutation primer directing the desired mutation and spanning the target mutation site was used for amplification. Following purification of this first PCR fragment and its direct use as a primer together with the alternate primer a second round of DNA amplification on the same plasmid DNA was performed. This now mutated fragment was purified again, restricted and ligated into the original parent plasmid followed by transformation directly into *E. coli* to obtain a recombinant clone [23].

PCR Products as Internal Controls

A rapid and highly versatile method for synthesis of internal DNA standards was described by Celi et al. [29] and an improved and more detailed protocol was published by Förster [30].

The principle is shown in Figure 1.2-2. A competitor fragment of defined size for a chosen PCR product is generated in two consecutive steps. Standard PCR primers (Pr1, Pr2) for the chosen sequence and an additional linker primer (Pr3) are needed. Following amplification of the DNA sequence with the outer primers the first reamplification step was performed by replacing primer 2 by a 3' linker primer (Pr3) carrying the original sequence of primer 2 on its 5' end. The second reamplification step of the fragment generated with primer 1 and 3 was performed with the original primers yielding a shortened fragment. Alternatively, the internal standard can be generated by a single step using primers 1 and 3 to synthesize a desired competitor in a single day [29]. Following purification and quantitation by conventional techniques (e.g. HPLC, see chapter 3.3 and 3.4) a known number of copies of this internal standard may be introduced into the sample and amplified with primers 1 and 2. With this

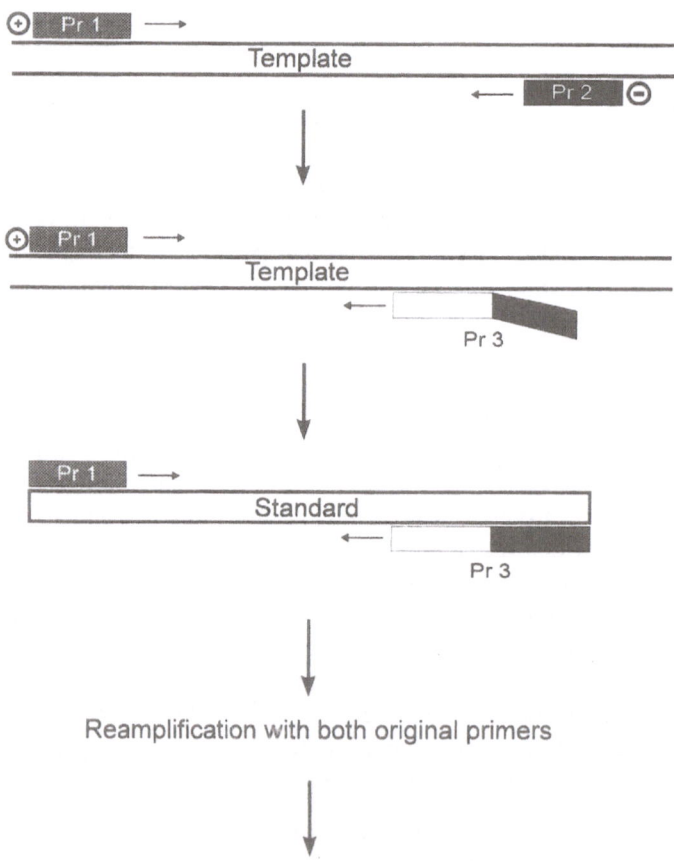

● Purification of standard fragment by electrophoresis or HPLC
● Calibration against a DNA standard of known concentration

Fig. 1.2-2. A simple method for generation of internal standards by PCR

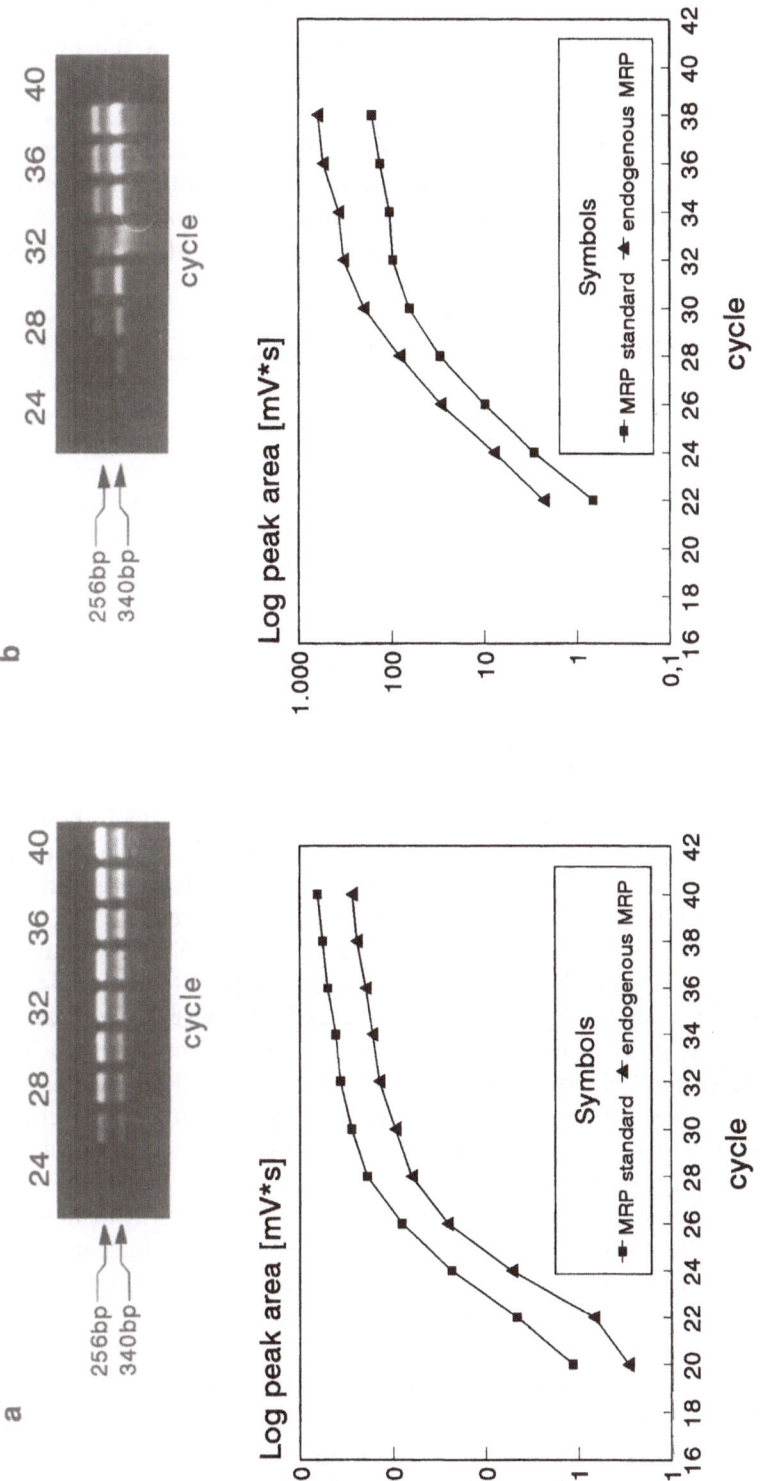

Fig. 1.2-3. Kinetic analysis of simultaneous amplification of MRP-cDNA and internal standard. Both standard and product accumulate in a parallel manner over the whole cycle range even the tested standard concentrations differ in one order of magnitude. **a** 2 μl of RT reaction product corresponding to cDNA synthesized from 100 ng of total RNA from the CCRF ADR5000 cell line [42]; 0.15 amol standard; **b** 100 ng cDNA, 0.015 amol standard per reaction tube

approach, two products are generated, one derived from the endogenous template, and another, some base pairs smaller derived from the internal standard (Figure 1.2-2).

The advantage of this method is the generation of internal standards which do not differ from the DNA sequence of the endogenous derived fragment (homologous competitors) allowing amplification with identical efficiencies for both products and quantitation at plateau conditions even when the used internal standard concentrations differ in one order of magnitude (Figure 1.2-3).

1.2.3 What Should be the Internal Standard of Choice?

For measuring absolute numbers of starting amounts of a mRNA of interest the internal standard of choice should fulfill a number of criteria:

- It must be amplified with the same efficiency as the natural mRNA target.
- The standard must control for RNA extraction and cDNA synthesis.
- The different PCR products must be distinguishable.
- It should produce a standard curve by generation of the ratio of internal control to target, rather than extrapolating against a standard curve.
- It must control for intratube variations at the level of cDNA synthesis and PCR amplification.
- Internal PCR standards should be exactly measurable and storable over a long time period.

All these criteria are best fulfilled by the use of RNA standards which show the highest possible similarity (homologous competitors) to the target sequence and added prior to reverse transcription to allow quantitation at plateau conditions. But some limitations of cRNA usage were reported by Guiffre et al. [25] who described problems to yield DNA-free cRNA. The often used DNase I treatment [5, 22, 25, 27] to purify the cRNA transcript from the template (e.g. plasmid) or an oligo (dT) chromatography [4] may be unsuccessful because of the presence of SP6 polymerase which is often used to synthesize cRNA seems to be sufficient to protect the vector DNA from degradation by DNase I in cRNA purified by standard procedures. This problem was overcome by centrifuging the cRNA preparation through a cesium chloride gradient followed by phenol/chloroform extraction of the transcription product. Another disadvantage was reported by Babu et al. [43] who found that sometimes the internal standard RNA competes with the target mRNA when both are present in disproportionate concentrations in the initial simultaneous RT reaction. These limitations were circumvented by the competitor DNA standard approach and quantitation of mRNA levels by calculating the RT efficiency.

As is generally known, RNA is very susceptible to ubiquitous RNases. Special efforts, e.g. exclusive use of DEPC treated solutions, RNase inhibitors, are therefore necessary to protect RNA standards from degradation. Since RNA extraction and reverse transcription often have rather viable yield [15] the inclusion of internal RNA as standard would not check for variations in RNA extraction if

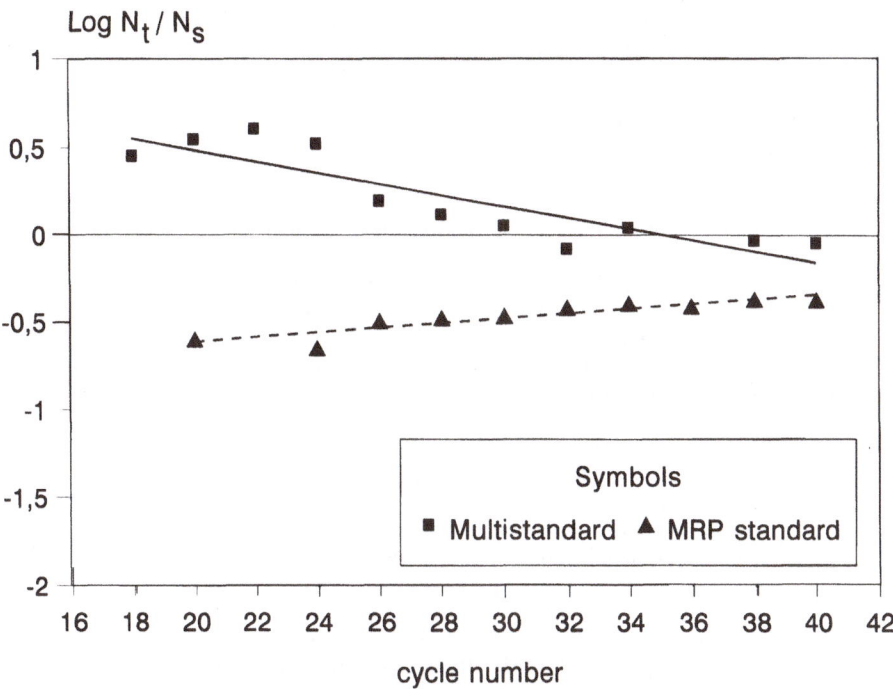

Fig. 1.2-4. Identification of efficiency differences between endogenous and standard molecules (data expressed in terms of the E_t/E_s ratio according to [21]) *Dashed line:* MRP standard generated by site-directed mutagenesis as described [29, 30], *Continous line:* Simultaneous amplification of mdr-1 cDNA and calibrated multistandard PCR fragments. Significant differences in amplification of standard and endogenous target result in non-horizontal E_t/E_s curves (*continous curve*). The efficiencies for standard and target amplification calculated by linear regression analysis were for the MRP standard 1.09 and 1.32 and for the multistandard 0.59 and 0.90 respectively

not added at the level of RNA preparation from the biological source [15]. If only comparative but not absolute results are required the use of an endogenous internal control such as the ubiquitous GAPDH mRNA has several advantages. This kind of internal control was described as being present in all cellular samples and processed under identical conditions. So an extensive series of reactions per sample is not necessary and, a synthetic internal standard does not need to be cloned or synthesized [10].

Although DNA cannot be utilized to check for the reverse transcription step we favor DNA standards because they are much more stable, can be stored frozen as stock solutions over a long period, may be precisely calibrated e.g. by HPLC (see chapter 3.3.) and allow additional contamination security if dUTP nucleotides are permanently incorporated in all synthesized PCR products. As shown in Figure 1.2-4, standards generated by site-directed mutagenesis according to the method described by Förster [30] (see above) may have some advantages over multistandards generated from synthetic genes. In contrast to the multistandard, such competitors may be amplified with the same efficiency as the endogenous target over the whole cycle range (Figure 1.2-4) allowing quantitation even with reactions driven to the plateau.

References

1. Siebert PD (1993) Quantitative RT-PCR. Methods & Applications; Book 3, Clontech Laboratories, Inc.
2. Taylor GR (1991) Polymerase chain reaction: basic principles and automation. In: McPherson MJ, Quirke P, Taylor GR (eds.): PCR: A practical approach. Practical approach series, Oxford University Press, New York, pp 1–14
3. Chelly J, Montarras D, Pinset C, Berwald-Netter Y, Kaplan J-C, Kahn A (1990) Quantitative estimation of minor mRNAs by cDNA-polymerase chain reaction. Eur J Biochem; 187:691–698
4. Wang AM., Doyle MV, Mark DF (1989) Quantitation of mRNA by the polymerase chain reaction. Proc Natl Acad Sci USA; 86:9717–9721
5. Bouaboula M, Legoux P, Pessegues B, Delpech B, Dumont X, Piechaczyk M, Casellas P, Shire D (1992) Standardization of mRNA titration using a polymerase chain reaction method involving co-amplification with a multispecific internal control. J Biol Chem; 267:21830–21838
6. Bremer S, Hoof T, Wilke M, Busche R, Scholte B, Riordan, JR, Mass G, Tümmler B (1992) Quantitative expression patterns of multidrug-resistance P-glycoprotein (MDR1) and differentially spliced cystic-fibrosis transmembrane-conductance regulator mRNA transcripts in human epithelia. Eur J Biochem; 206:137–149
7. Katz ED, DiCesare JL, Picozza E, Enderson MS (1993) General aspects of PCR quantitation. Amplifications; 10:7–8
8. Kageyama Y, Koide Y, Miyamoto S, Inoue T, Yoshida TO (1994) The biased V_g gene usage in the synovial fluid of patients with rheumatoid arthritis. Eur J Immunol; 24:1122–1129
9. Sardelli AD (1993) Plateau effect – Understanding PCR limitations. Amplifications; 9: 1–5
10. Wong H, Anderson WD, Cheng T, Riabowol KT (1994) Monitoring mRNA expression by polymerase chain reaction: the "primer-dropping" method. Anal Biochem; 223: 251–258
11. Erlich HA, Gelfand D, Sninsky JJ (1991) Recent advances in the polymerase chain reaction. Science; 252:1643–1651
12. Lönn U, Lönn S, Nylen U, Stenkvist B (1992) Appearance and detection of multiple copies of the mdr-1 gene in clinical samples of mammary carcinoma. Int J Cancer; 51:682–686
13. Kinoshita T, Imamura J, Nagai H, Shimotohno K (1992) Quantification of gene expression over a wide range by the polymerase chain reaction. Anal Biochem; 206:231–235
14. Zhou J, Hoffman EP (1994) Pathophysiology of sodium channelopathies. J Biol Chem; 269:18563–18571
15. Pannetier C, Delassus S, Darche S, Saucier C, Kourilsky P (1993) Quantitative titration of nucleic acids by enzymatic amplification reactions run to saturation. Nucl Acids Res; 21:577–583
16. Murphy LD, Herzog CE, Rudick JB, Fojo AT, Bates SE (1990) Use of polymerase chain reaction in the quantitation of mdr-1 gene expression. Biochemistry; 29:10351–10356
17. Lyttelton MP, Hart S, Ganeshaguru K, Hoffbrand AV, Mehta AB (1994) Quantitation of multidrug resistant MDR1 transcript in acute myeloid leukemia by non-isotopic quantitative cDNA-polymerase chain reaction. Br J Haematol; 86:540–546
18. Hoof T, Riordan JR, Tümmler B (1991) Quantitation of mdr1 transcript by PCR: a tool for monitoring drug resistance in cancer chemotherapy. In: Rolfs A, Schumacher HC, Marx P (eds.) PCR topics. Springer, Berlin, Heidelberg, New York, pp. 217–220
19. Dostal DE, Rothblum KN, Baker KM (1994) An improved method for absolute quantification of mRNA using multiplex polymerase chain reaction: determination of renin and angiotensin mRNA levels in various tissues. Anal Biochem; 223:239–250

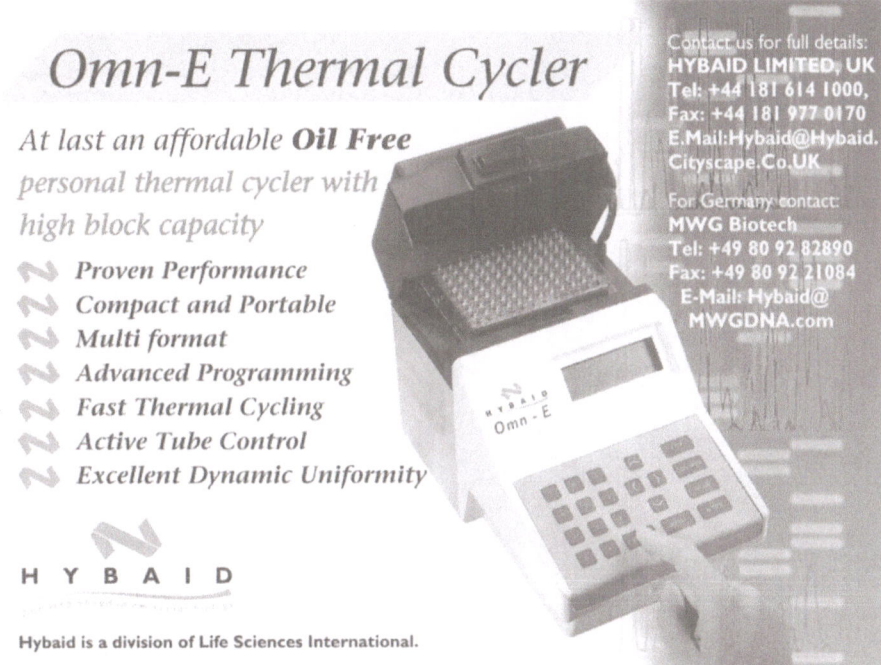

20. Dukas K, Sarfati P, Vaysse N, Pradayrol L (1993) Quantitation of changes in the expression of multiple genes by simultaneous polymerase chain reaction. Anal Biochem; 215: 66–72

21. Scheuermann RH, Bauer SR (1993) Polymerase chain reaction-based mRNA quantification using an internal standard: Analysis of oncogene expression. Methods Enzymol; 218:446–473

22. Guyon T, Levasseur P, Truffault F, Cottin C, Gaud C, Berrih-Aknin S (1994) Regulation of acetylcholine receptor alpha subunit variants in human myasthenia gravis. Quantification of steady-state levels of messenger RNA in muscle biopsy using the polymerase chain reaction. J Clin Invest; 94:16–24

23. Chen B, Przybyloa AE (1994) An efficient site-directed mutagenesis method based on PCR. BioTechniques; 17:657–659

24. Futscher BW, Blake LL, Gerlach JH, Grogan TM, Dalton WS (1993) Quantitative polymerase chain reaction analysis of mdr1 mRNA in multiple myeloma cell lines and clinical specimens. Anal Biochem; 213:414–421

25. Guiffre A, Atkinson K, Kearney P (1993) A quantitative polymerase chain reaction assay for interleukin 5 messenger RNA. Anal Biochem; 212:50–57

26. Nagano M, Kelly PA (1994) Tissue distribution and regulation of rat prolactin receptor gene expression. J Biol Chem; 269:13337–13345

27. Prendergast JA, Helgason CD, Bleackley RC (1992) Quantitative polymerase chain reaction analysis of cytotoxic cell proteinase gene transcripts in T cells. J Biol Chem; 267: 5090–5095

28. Morales MJ, Gottlieb DI (1993) A polymerase chain reaction-based method for detection and quantification of reporter gene expression in transient transfection assays. Anal Biochem; 210:188–194

29. Celi FS, Zenilman ME, Shuldiner AR (1993) A rapid and versatile method to synthesize internal standards for competitive PCR. Nucl Acids Res; 21:1047

30. Förster E (1994) An improved general method to generate internal standards for competitive PCR. BioTechniques; 16:18–20

31. Wadhwani KC, Fukuyama R, Giordano T, Rapoport SI, Chandrasekaran K (1993) Quantitative reverse transcriptase-polymerase chain reaction of glucose transporter 1 mRNA levels in rat brain microvessels. Anal Biochem; 215:134–141

32. Rhoer-Moja S, Cohen-Haguenauer O, Jouve C, Healy J-C, Vindimian M (1993) Detection of quantitative polymerase chain reaction products by hybridization on magnetic support with ^{125}I-radiolabeled probes: Quantification of c-myc copy numbers. Anal Biochem; 213: 12–18

33. Yokoi H, Natsuyama S, Iwai M, Noda Y, Mori T, Mori KJ, Fujita K, Nakayama H, Fujita J (1993) Non-radioisotopic quantitative RT-PCR to detect changes in mRNA levels during early mouse embryo development. Biochem Biophys Res Comm; 195:769–775

34. Caroff N, Della Brunna R, Philippe J, Corvol P, Pinet F (1993) Regulation of human renin secretion and renin transcription by quantitative PCR in cultured chorionic cells: Synergistic effect of cyclic AMP and protein kinase C. Biochem Biophys Res Comm; 193: 1332–1338

35. Kovacs DM, Kaplan BB (1992) Discordant estimates of heterologous promoter activity as determined by reporter gene mRNA levels and enzyme activity. Biochem Biophys Res Comm; 189:912–918

36. Jackson DP, Hayden JD, Quirke P (1991) Extraction of nucleic acid from fresh and archival material. In: McPherson MJ, Quirke P, Taylor GR (eds.): PCR: A practical approach. Practical approach series, Oxford University Press, New York, pp. 29–50

37. Bebee RL, Thornton CG, Hartley JL, Rashtchian, A (1992) Contamination-free polymerase chain reaction: endonuclease cleavage and cloning of dU-PCR products. Focus; 14:53–56

38. Shire D and the Editorial Staff of E.C.N. (1993) An invitation to an open exchange of reagents and information useful for the measurement of cytokine mRNA levels by PCR. Eur Cytokine Netw; 4:161–162

39. Lundeberg J, Wahlberg J, Uhlen M (1991) Rapid colorimetric quantification of PCR amplified DNA. BioTechniques; 10:68–75

40. Zhou L, Pang J, Munroe DG, Lau C (1993) A human retinoic acid receptor gamma isoform is homologous to the murine retinoic acid receptor gamma 7. Nucl Acids Res; 21:2520

41. Mashiyama S, Murakami Y, Yoshimoto T, Sekiya T, Hayashi K (1991) Detection of p53 gene mutations in human brain tumors by single-strand conformation polymorphism analysis of PCR products. Oncogene; 6:1313–1318
42. Gekeler V, Weger S, Probst H (1990) MDR1/P-glycoprotein gene segments analyzed from various human leukemic cell lines exhibiting different multidrug resistance profiles. Biochem Biophys Res Comm; 169:796–802
43. Babu JS, Kanangat S, Rouse BT (1993) Limitations and modifications of quantitative polymerase chain reaction. Application to measurement of multiple mRNAs present in small amounts of sample RNA. J Immunol Methods; 165:207–216
44. Erlich HA (ed.) (1989) PCR technology. Principles and applications for DNA amplification. MacMillan Publishers, London
45. Innis MA, Gelfand DH, Sninsky JJ, White TJ (eds.) (1990) PCR protocols. A guide to methods and applications. Academic Press Inc., Harcourt Brace, Jovanovich Publishers, San Diego
46. De Kant E, Rochlitz CF, Herrmann R (1994) Gene expression analysis by a competitive and differential PCR with antisense competitors. BioTechniques; 17:934–942

Chapter 1.3

Cloning of Short DNA Fragments and In Vitro Transcription to Generate RNA Standards

D. LASSNER

1.3.1 Theoretical Background

Quantitation of mRNA by polymerase chain reaction requires valuation of the possible modification of native mRNA molecule through cDNA synthesis (see chapter 2.2), following amplification (see chapters 2.3 and 2.4) and quantitation of generated PCR products by several methods (see chapters 2.5; 3.1; 3.3; 3.4). Addition of an internal RNA standard during extraction step in RNA isolation protocol is the most reliable check of processing of the initial RNA molecule in complete RT-PCR. The use of DNA standards can also be successful (see also chapter 1.2). A disadvantage of using a DNA standard is the different transcription rate in cDNA synthesis.

Generation of standards was very laborious before PCR technology reached molecular biology laboratories. For the reproducible synthesis of a desired standard, cloning selected nucleic acid fragment into a vector such as plasmids or phagemids is recommended. Searching for and inserting a new DNA fragment without the use of PCR is very difficult and good practical experience in molecular cloning is obligatory.

DNA cloning consists of three steps:

- Enrichment of DNA fragments selected for ligation and cloning
- Efficient cloning into a suitable vector
- Multiplying the cloned fragment

By PCR selection and enrichment of any fragment from a pool of millions of different sequences is possible. Synthesis of the necessary amount of a definite fragment is the first step of cloning. Specificity of selection is given by an applied primer pair in amplification.

Three methods are favored for ligation of PCR products [1]. The first method is ligation of blunt-ended PCR products into a vector which is digested with a restriction enzyme that cuts the vector which also becomes blunt-ended. This method of cloning is very difficult and the amplification product is randomly inserted in the cloning site. More successful are two other methods of cloning which use the specificity of the PCR process.

The 5'-terminus of a primer is not essential for specific annealing and the resulting extension of this in subsequent DNA synthesis by Taq polymerase. New restriction sites can be incorporated in a PCR product by modification of 5'-termini of one or both primers used. Throughout amplification, sufficient

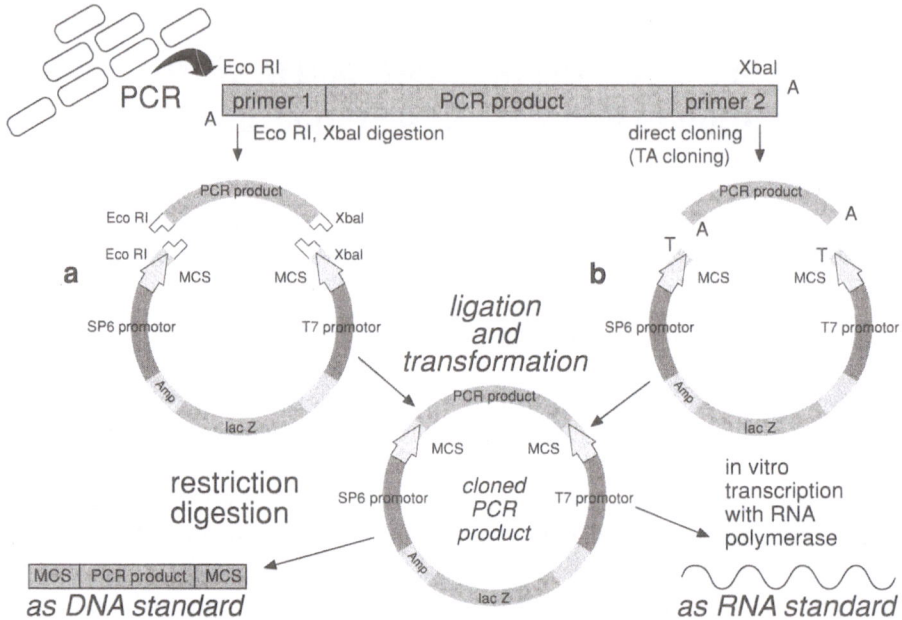

Fig. 1.3-1. Cloning of PCR products: **a** Unique restriction sites were incorporated into previously amplified PCR products by using specific primers. Following digestion with restriction endonucleases the sticky ended fragments were ligated into prepared plasmids. **b** T/A cloning. Single adenosine overhang at 3'-termini of PCR products allows simple ligation into a vector with single thymidine overhang

DNA fragments of a specific sequence flanked by two specific restriction sites are accumulated. After restriction enzyme digestion of the amplified sample, the selected sequence can be cloned into the vector which is digested by the same restriction enzymes. This method offers the possibility of directional cloning (see Figure 1.3-1, a).

Taq DNA polymerase creates a single adenosine overhang at the 3'-termini of newly synthesized PCR products. This property is exploited in T/A cloning [2]. The amplification product generated is cloned by a conventional ligation reaction into a special vector which has a single thymidine overhang (see Figure 1.3-1, b). The universal use of this cloning overcomes the disadvantage of non-directed insertion of DNA fragment.

Plasmids were transferred into bacteria such as *Escherichia coli* and after growing on culture media the transferred vector can be prepared with high yields.

An ideal standard has the same sequence, same length and primer binding sites as the native sequence. Therefore it is optimal to clone a PCR product which is generated by amplification of a native DNA/cDNA fragment.

Cloning should be performed with a vector with a universal multicloning site (e.g. a multitude of restriction sites symmetrically and asymmetrically distributed, two blunt-end generating endonucleases), promoter sites for RNA polymerases to perform in vitro transcription for designing cRNA standards and include a lac Z operon for selecting positive clones by blue/white screening.

1.3.2 Experimental Procedures

Cloning of PCR product can be done in many different ways. Here is a possible procedure for cloning a PCR product up to the generation of cRNA standards. Following steps are described separately:

- The T/A cloning method
- Transformation of *E. coli*
- Selection of positive colonies
- Plasmid preparation
- In vitro transcription with T7 RNA polymerase

1.3.2.1 T/A Cloning Procedure

The "TA Cloning Kit" (Invitrogen) is used for T/A cloning and the subsequent transformation. With this kit, all chemicals, media and even competent cells for ligation and efficient transformation are provided.

The T/A cloning system provides a quick, one-step cloning strategy for the direct insertion of a PCR product into a plasmid vector. The procedure eliminates any enzymatic modifications of the PCR product such as Klenow or T4 polymerase treatment to create blunt ends, and it does not require the use of PCR primers which contain restriction sites.

Generally, 100 ng of DNA are sufficient to serve as template for PCR. If amplifying cDNA, the amount needed will depend on the relative abundance of the message of interest in your mRNA population and the reaction mixture. Perform the amplification reaction in a 50 µl volume as described in chapters 2.3 and 2.4. For optimal ligation efficiencies, the use of no more than 30 PCR cycles is recommended. Analyze the PCR products by agarose gel electrophoresis through a 0.8–1.5% agarose gel (see Sect. 2.3.2.2).

T/A Cloning Ligation

There is no need to clean or purify the PCR product after the final amplification cycle. For optimal ligation efficiencies, use only freshly prepared PCR products. The single 3' A-overhangs on the PCR products will degrade over time, reducing ligation efficiency. Ligations with the pCRII vector should be set up as a 1:1 to 1:3 molar ratio of the vector to the PCR product. After running a few microliters of the completed PCR reaction product on a gel to estimate its concentration, use the formula below to determine the amount of PCR product to be ligated with 50 ng of pCRII vector. Then set up the ligation reaction as outlined below:

$$X \text{ ng of PCR product} = \frac{(Y \text{ bp PCR product}) * (50 \text{ ng pCR II vector})}{(\text{size in bp of the pCR II vector})}$$

where: X ng is the amount of PCR product of Y base pairs to be ligated for a 1:1 molar ratio.

The pCR II vector is 3932 bp in length. Use 3 * X for a 1:3 molar ratio.

Materials:

1) *Water bath (12 °C)*
2) *10 Ligation Buffer (provided with the kit)*
3) *Resuspended pCRII vector (25 ng/µl, provided with the kit)*
4) *T4 DNA ligase (provided with the kit)*
5) *PCR product (see chapter 2.3 or 2.4)*
6) *Sterile water*
7) *Ice bucket with ice*
8) *Mastertips and Biomaster (Eppendorf)*
9) *Eurotips and Varipettes 4810 (Eppendorf)*

Protocol 1:   Direct Ligation of the Generated PCR Product

Pipette the following solutions to perform one ligation reaction (all steps on ice):

- Sterile Water   5 µl
- 10 × Ligation Buffer   1 µl
- Resuspended pCRII vector (25 ng/µl)   2 µl
- PCR product   1 µl
- T4 DNA Ligase   1 µl

Incubate ligation reaction at 12 °C for a minimum of 4 hours (preferably over-night). Ligation at higher or lower temperatures will diminish the ligation efficiency.

T/A Cloning Transformation

This protocol represents a general way to transform competent cells. With the "TA Cloning" kit, ready-to-use competent cells are provided. Store these cells at –70 °C.

Be extremely gentle when working with competent cells because they are highly sensitive to changes in temperature or mechanical lysis caused by pipetting. Transformation should be started immediately following the thawing of the cells on ice and mixing should be done by swirling or tapping the tube gently, not by pipetting.

Sterile conditions for growing of bacteria are essential. For details see the relevant literature [3,4].

Materials:

1) *42 °C water bath*
2) *37 °C incubator*
3) *10 cm diameter LB agar plates with appropriate antibiotic (50 μg kanamycin/ml agar)*
4) *X-Gal stock solution (40 mg/ml DMSO)*
5) *Ice bucket with ice.*
6) *SOC medium (provided within the kit)*
7) *0.5 mol/l β-mercaptoethanol (provided within the kit)*
8) *OneShot competent cells (E.coli strain)*
9) *Mastertips and Biomaster (Eppendorf)*
10) *Eurotips and Varipettes 4810 (Eppendorf)*

Protocol 2: Transformation of Competent Cells

1. Heat a water bath to 42 °C.
2. Warm up one vial of SOC medium to room temperature.
3. Place an appropriate number of 10 cm diameter LB agar plates with antibiotic in an incubator at 37 °C to remove excess moisture. Use two plates for each ligation/transformation reaction.
4. Use a test tube rack that will hold all of the transformation tubes so that they can all be put into the 42 °C water bath at once.
5. Spin the vials containing the ligation reactions briefly and place them on ice.
6. On ice, thaw 0.5 mol/l β-mercaptoethanol and one 50 μl vial of frozen OneShot competent cells for each ligation/transformation.
7. Pipette 2 μl of the 0.5 mol/l β-mercaptoethanol into each vial of the competent cells and mix by tapping gently.

>>Caution: Do not mix by pipetting up and down !<<

8. Pipette 1 μl of each T/A Cloning ligation reaction. Store the remaining ligation mixture(s) at −20 °C.
9. Incubate the vials on ice for 30 min.
10. Incubate for exactly 30 s in the 42 °C water bath. Do not mix.
11. Remove the vials from the 42 °C water bath and immediately place on ice for 2 min.
12. Add 450 μl of pre-warmed SOC medium to each vial (SOC is a rich medium and good sterile technique must be used to prevent contamination).

13. Place the vials in a microcentrifuge rack and fasten with Parafilm to avoid loss of the vials. Shake the vials at 37 °C for exactly 1 h at 225 rpm in a rotary shaking incubator.

14. While the incubation is going on, prepare an L-shaped glass spreader by first flaming the end of a glass Pasteur pipette shut, then bending the thin portion of the pipette in the flame. Prepare LB agar plates containing kanamycin (50 µg/ml) or ampicillin (50 µg/ml) by spreading 25 µl of X-Gal on top of agar with the glass spreader. Let the X-Gal diffuse into the agar for approximately 1 h.

15. Place the vials with the transformed cells on ice.

16. Spread 25 µl and 100 µl from each transformation vial on separate, labeled LB agar plates containing antibiotic and X-Gal.

17. Invert the plates and place them in a 37 °C incubator overnight.

Note: In some cases, a longer incubation (up to 40 h) is required to allow the blue color to develop in the background colonies.

18. Pick the white colonies for plasmid isolation and restriction analysis, PCR, or sequencing.

Note: InvaF' competent cells of TA Cloning kit are derivatives of an *E. coli* K12 strain with genotype F'endA1 recA1 hsdR17 (r-K,m + K) l- supE44 thi-1 gyrA96 relA1 F80DlacDM15D(lacZYA-argF)U169 deoR.

Trouble Shooting:

When cloning PCR products of 500 bp or less, light blue, rather than white, colonies may appear. The reading frame of the lacZ gene may not have been disrupted in these colonies. Treat these light blue colonies as if they were white because they may contain the insert. Spreading of 100 µl from each transformation can lead to a filled agar plate. IPTG is not required since the OneShot cells do not express the lac Iq repressor (for details see [4]). Typically, plating out 25 µl of transformed cells yields ~50 colonies.

We have cloned a 300 bp DNA fragment and obtained distinctly white colonies. In contrast, it has been difficult to clone a 1700 bp-fragment of mdr-1 mRNA.

1.3.2.2 Minipreparation of Plasmid DNA

The first hint to successful transformation of *E. coli* cells with cloned plasmid is the occurrence of white colonies on the LB agar plate. Cloning of the desired PCR fragment in the pCR II vector is statistically predominate. Positive colonies of transformed *E. coli* cells which contain the plasmid with the inserted PCR product may be detected after minipreparation of plasmid DNA.

The ideal method for the isolation of single positive colonies is called "streaking". Each colony identified as "white" is streaked on a separate plate as described in detail [3]. From each streaked plate, white colonies are picked and a liquid culture of each is made. After growing overnight plasmids are purified from cells by alkaline lysis and several purification steps [5].

Materials:

1) *LB agar plate containing kanamycin (50 µg antibiotic/ml agar)*
2) *LB medium for overnight culture with appropriate antibiotic*
3) *NaOH/SDS solution (0.2 mol/l NaOH, 1% (v/v) SDS)*
4) *Potassium acetate solution (see appendix)*
5) *100% Ethanol*
6) *70% (v/v) Ethanol*
7) *X-Gal (40 mg/ml DMSO)*
8) *RNase, DNase-free (Boehringer Mannheim)*
9) *TE buffer (see appendix)*
10) *Glass spreader*
11) *Inoculating loop*
12) *Incubator (37 °C) with shaking table*
13) *Sterile toothpicks*
14) *Ice bucket with ice*
15) *Vortexer*
16) *Eppendorf reaction tubes (Eppendorf)*
17) *Eurotips and Varipettes 4810 (Eppendorf)*
18) *Heto vacuum concentrator (Heto, Scandinavia)*

Protocol 3: Minipreparation of Plasmid DNA

1. Prepare LB agar plates with appropriate antibiotic by spreading X-Gal as described above.
2. Use two LB agar plates for 10 white colonies obtained from transformation.
3. Pick each white colony with one sterile toothpick on two separate agar plates. Mark the position of identical colonies on both plates. Pick up to 10 colonies from each plate to get finally two identical plates.
4. Incubate the plates in an incubator (37 °C) overnight. Next day check that all colonies are visible on both plates. Store one plate at +4 °C and use the other for plasmid minipreparation.
5. Inoculate 5 ml LB medium with one colony from the agar plate. Grow to saturation overnight by shaking at 225 rpm at 37 °C in an incubator with a shaking plate.
6. Transfer 1.5 ml of cells to a sterile Eppendorf reaction tube and spin 20 s in a microcentrifuge.
7. Remove the supernatant and resuspend pellet in 100 µl TE buffer and incubate for 5 min at room temperature.

8. Add 200 µl NaOH/SDS solution, tap the tube with the finger and place on ice for 5 min.
9. After addition of 150 µl potassium acetate solution vortex the tube at highest speed and place again on ice for 5 min.
10. Centrifuge for 1 min and transfer the supernatant to a new Eppendorf reaction tube.
11. Add 0.9 ml of 100% ethanol and allow to incubate at room temperature for 2 min.
12. Centrifuge 1 min and remove supernatant. Add 1 ml of 70% ethanol and repeat centrifugation.
13. Dry pellet under vacuum and resuspend in 100 µl TE buffer.
14. Pipette 1 µl of RNase to the prepared sample and incubate for 1 hour at 37 °C to eliminate all remaining RNA.

This original protocol of Birnboim and Doly [5] is the basis of many commercially available miniprep kits. Pure, RNA-free plasmids can be prepared in less than one hour using a "QIAprep Spin Plasmid Kit" (Qiagen).

The amount of purified plasmid DNA is determined by UV spectroscopy (see Sect. 2.1.6) and examined by agarose gel electrophoresis after restriction digestion as described below.

1.3.2.3 Digestion of Isolated Plasmid DNA with Restriction Endonucleases

Most of the commercially available plasmids are fitted out with multiple cloning sites (MCS). In general, DNA fragments which we want to clone are inserted between the "sticky ends" of a multicloning site generated by digestion of the plasmid preparation with corresponding restriction enzymes. In the same way cloned DNA fragments can be removed from the plasmid and distinguished following agarose gel electrophoresis from the linearized plasmid.

The structure of plasmid pCR II containing an insert synthesized by PCR is schematically demonstrated in Figure 1.3-1.

A mdr-1 mRNA derived fragment is inserted into the EcoRI restriction site of the pCR II vector. Digestion of the plasmid containing the insert by this enzyme and analysis on agarose gel (see Sect. 2.3.2.2) results in two different bands: one of 3932 bp (vector pCR II) and one of 289 bp (cloned fragment). If the vector is digested with Xba I, which recognizes only one intramolecular site, only one single band representing the linearized molecule is obvious.

The MCS of vector pCR II is, in addition, flanked by the two different viral RNA polymerase transcription promoters SP6 and T7. This property may be exploited for high-level in vitro transcription of one of the two strands from the inserted PCR fragment derived by simple selection of the respective RNA polymerase [6].

A fragmentation of the construct into 31 restriction fragments may be achieved by digestion with endonuclease Hha I. The 743 bp fragment obtained includes the full multiple cloning site, both promoter sequences and the cloned mdr-1 PCR fragment.

Materials:

1) *Hha I (10 U/μl, Pharmacia)*
2) *10 × One-Phor-All buffer Plus (100 mmol/l Tris/HCl; pH 7.5; 100 mmol/l Mg-acetate, 500 mmol/l K-acetate) (Pharmacia)*
3) *Plasmid preparation*
4) *Lambda DNA-EcoRI & HindIII digest (AGS)*
5) *DEPC-H₂O*
6) *Thermomixer (Eppendorf)*
7) *Mastertips and Biomaster (Eppendorf)*
8) *Eurotips and Varipettes 4810 (Eppendorf)*

Pipette the following solutions to prepare one restriction digestion assay of plasmid preparation (50 μl):

- 10 × One-Phor-All buffer Plus   5 μl
- DEPC-H₂O   24 μl
- Plasmid (about 1 μg)   20 μl
- HhaI   1 μl

Incubate at 37 °C for 1 h and inactivate enzyme by additional incubation at 65 °C for 15 min. Load 10 μl of restriction digest onto an agarose gel (0.8 – 1.5 %, w/v) and examine the results as described in Sect. 2.3.2.2. Use a 100 base-pair ladder (Pharmacia) as length size marker to examine Hha I digest composition. If only the cleaved plasmid is separated use a λ-phage DNA-digest (EcoRI/HindIII digested λ-DNA, AGS) as size marker.

Digestion with other endonucleases may be performed in a very similar way. Most restriction enzymes show a temperature optimum at about 37 °C and can be easily inactivated at 65 °C for 15 min.

If the insert is needed for further application carry out cleavage (e.g. with Eco RI) and concentrate the sample by vacuum drying and dissolving in 10 μl DEPC-H₂0 or buffer. Load the whole concentrated digest on agarose gel. This way is preferable when the cloned fragment is very short (less than 500 bp).

For primary check of the accuracy of the cloned fragment, purify an aliquot of the digested sample by phenol-chloroform-isoamyl alcohol extraction as mentioned below (Sect. 1.3.2.4). Dissolve the extracted material and perform a reamplification with the corresponding primers. If a desired PCR product is obtained the right sequence was cloned.

1.3.2.4 In Vitro Transcription of Cloned Fragments by T7 RNA Polymerase

In vitro transcription means the generation of single stranded RNA from a cloned DNA template using a DNA-dependent RNA polymerase that is highly specific for its promoter sequence. The promoter represents the starting point of the very rapid and extremely processive RNA synthesis. If a cloned cDNA fragment is in vitro transcribed into RNA the resulting transcripts are called cRNA ("copy-RNA").

Previous linearization of purified plasmid by restriction digestion increases the yield of in vitro transcribed cRNA.

After in vitro transcription, the remaining DNA template must be hydrolyzed with 10 U DNase (RNase-free) at 37 °C for 1 h. If the amount of DNA introduced into the assay is too high the DNase treatment may be insufficient. It is therefore advisable to generate at first a target fragment consisting of the cloned PCR product flanked by one remaining promoter sequence followed by separation on agarose gel and subsequent extraction from the respective agarose slices (see Sect. 2.6.2.2). The purified fragment is now ready to use for in vitro transcription and subsequent DNase digestion without the problems mentioned.

In the case of cloned mdr-1 PCR product a Hha I digestion of pCR II vector (Sect. 1.3.2.3) is performed and the 739 bp fragment is separated from the agarose gel as described in Sect. 2.6.2.2.

In vitro transcription requires an absolutely pure DNA preparation. Therefore it is sometimes necessary to perform a subsequent phenol-chloroform-isoamyl alcohol extraction as described below.

Materials:

1) *Buffer-saturated phenol (RotiPhenol, Roth)*
2) *Chloroform-isoamyl alcohol (24:1, v/v) mixture (–20 °C)*
3) *Na-acetate solution (2.5 mol/l Na-acetate)*
4) *Ethanol (100%, –20 °C)*
5) *Ethanol (75%, –20 °C)*
6) *Vortexer*
7) *Eppendorf reaction tube (Eppendorf)*
8) *Eurotips and Varipettes 4810 (Eppendorf)*
9) *Centrifuge 5415 C (Eppendorf)*
10) *Heto vacuum concentrator (Heto, Scandinavia)*

Protocol 4: Extraction of Pure, Protein Free Nucleic Acid Preparations by Phenol-Chloroform-Isoamyl Alcohol

1. Dissolve gel-purified DNA pellet in 100 µl DEPC-treated water (see Sect. 2.6.2.2).
2. Add a half volume of buffer-saturated phenol (50 µl) and a half volume of chloroform: isoamyl alcohol (24:1).
3. Mix thoroughly by vortexing and centrifuge the solution at 4000 rpm for 5 min.
4. Transfer the upper phase (about 100 µl) carefully to a new Eppendorf reaction tube.
5. Add one volume of chloroform-isoamyl alcohol (100 µl) and vortex thoroughly again.
6. Centrifuge the solution at 4000 rpm for 5 min and transfer the upper phase to a new Eppendorf tube.

7. Add 1/10 volume Na-acetate solution (10 µl) and 2.5 vol. of ice-cold 100% ethanol, precipitate nucleic acid overnight at – 20 °C or lower.
8. Centrifuge the tube at 12000 rpm for 15 min. Remove the supernatant immediately by inverting the tube over a paper towel to remove alcohol completely.
9. Wash the pellet obtained (could be invisible!) by addition of 100 µl of 75% ethanol and centrifuge the tube at maximum speed for 15 min.
10. Remove alcohol as described above and dry the pellet under vacuum.

Dried pellets are used for the in vitro transcription assay.

Materials:

1) *T7 RNA polymerase (20 U/µl, Boehringer Mannheim)*
2) *10 × Transcription buffer (0.4 mmol/l Tris/HCl; pH 8.0, 60 mmol/l MgCl₂, 100 mmol/l DTT, 20 mmol/l spermidine) (Boehringer, Mannheim)*
3) *NTPs (10 mmol/l of each ribonucleotide ATP, CTP, GTP, UTP, Pharmacia)*
4) *RNase inhibitor (20 U/µl, AGS)*
5) *DEPC-H₂0*
6) *DNase, RNase-free (Boehringer)*
7) *Thermomixer (Eppendorf)*
8) *Mastertips and Biomaster (Eppendorf)*
9) *Eurotips and Varipettes 4810 (Eppendorf)*

Protocol 5:   In Vitro Transcription of DNA Fragments with T7 RNA Polymerase

Add the following solutions using a Biomaster pipettor and Mastertips to the vacuum dried DNA pellet.
    Prepare a mastermix and transfer 20 µl of the mixture to the tubes containing the DNA target.
    Proceed as follows (pipetting scheme for one assay, respectively):

- DEPC-H₂O   12.5 µl
- 10 transcription buffer   2.0 µl
- ATP   1.0 µl
- CTP   1.0 µl
- GTP   1.0 µl
- UTP   1.0 µl
- RNase inhibitor   0.5 µl
- T7 RNA polymerase   1.0 µl

1. Incubate the completed in vitro transcription assay 2 h at 37 °C using a thermomixer and slight rotating.
2. Add 1 µl DNase (RNase-free) to each reaction mix and incubate for one hour at 37 °C. This step eliminates plasmid DNA in the transcription sample. Only freshly synthesized RNA remains.

3. Add 80 μl DEPC-treated water and perform extraction by phenol-chloroform-isoamyl alcohol (see above) to remove proteins and contaminating salt residues.
4. Dissolve the pellet in DEPC-treated water and estimate nucleic acid (RNA) content as described in Sect. 2.1.6. The RNA can now be applied for quantitation by using external standards (see chapter 3.1). Dilute the RNA sample with a solution of carrier RNA (tRNA of *E. coli*, Boehringer Mannheim) to increase stability of specific standard during storage at –20 °C (or long-time storage at –70 °C).

These RNA standards can be reverse transcribed accordingly to the RT reaction protocols (chapter 2.2) and simultaneously, with the endogenous transcript, amplified by PCR. Check the generation of cRNA standards very carefully. Check the standards if possible also by Northern blotting (chapter 2.6).

Trouble Shooting:

In this multiple step procedure, there are many possible sources of error. For cloning protocols, for more details see [3, 4, 6]. A problem for further quantitation is the remaining plasmid DNA in the cRNA standard solution. Remove all RNA by digestion of the standard solution with RNase (DNase free) and perform an amplification with specific primers. The DNase digestion (see step 2) of the in vitro transcription sample will not be complete if a PCR product is present [7].

References

1. Costa GL, Sanchez TR, Weiner MP (1994) pCR-Script Direct SK(+) vector for directional cloning of blunt-ended PCR products. Strategies; 7:5–7
2. Clark JM (1988) Novel non-templated nucleotide addition reaction catalyzed by procaryotic and eucaryotic DNA polymerases. Nucl Acids Res; 16:9677–9686
3. Ausubel FM (ed.) (1990) Current protocols in molecular biology. Greene Publishing Associates Inc. and John Wiley & Sons Inc., New York
4. Sambrook J, Fritsch EF, Maniatis T (eds.) (1989) Molecular Cloning: A laboratory manual , Cold Spring Harbor University Press, Cold Spring Harbor
5. Birnboim HC, Doly J (1979) A rapid alkaline extraction procedure for screening recombinant plasmid DNA. Nucl Acids Res; 7:1513–1523
6. Perbal B (ed.) (1988) A practical guide to molecular cloning. John Wiley & Sons Inc., New York
7. Guiffre A, Atkinson K, Kearney P (1993) A quantitative polymerase chain reaction assay for interleukin 5 mRNA. Anal Biochem; 212:50–57

Chapter 1.4

Direct Non-Isotopic Sequencing of PCR Products or Standards

B. Thamm

1.4.1 Theoretical Aspects

The direct sequencing method represents an efficient and uncomplicated technique to obtain the exact sequence information of PCR [1] products or synthetic PCR standards.

Since the introduction of the chain termination DNA sequencing method by Sanger et al. [2] a variety of technical modifications have been developed to improve the efficiency of sequencing reactions and to adapt them to the analysis of specific templates.

The basic principle of the method consists of the DNA polymerase catalyzed in vitro synthesis of populations of DNA strands complementary to the template DNA and differing in length by one nucleotide. The synthesis is initiated at a specific primer annealing site and terminated by the incorporation of a nucleotide analog that lacks the 3-OH group necessary for DNA chain elongation (2',3'-dideoxynucleoside 5'-triphosphates/ddNTPs). Optimal mixtures of dNTPs (providing elongation) and one of the four ddNTPs (stopping elongation) will result in the generation of DNA chains determined in length by the locations of the complementary base of this particular ddNTP in the template. Four separate reactions, each containing a different ddNTP, will give the complete sequence information after high-resolution gel electrophoresis.

The problem of sequencing short double-stranded DNA such as PCR products is the tendency of the templates to reanneal. To prevent renaturation we use two alternative sequencing strategies: The thermal cycling DNA sequencing utilizing a thermocycling apparatus and a solid phase DNA sequencing technique.

Cyclic sequencing yields a linear amplification of the template DNA, reducing the amount of template required to achieve a detectable sequence ladder. The high temperatures employed during each denaturation cycle help to circumvent the problems associated with rapid reannealing; the high annealing temperatures increase the stringency of primer hybridization. Sequencing grade Taq DNA polymerase (Promega) produces a sufficient uniform band intensity, low background and a high degree of accuracy.

The solid phase approach using Dynabeads (Dynal) as a magnetizable solid phase for the capture of PCR products yields an immobilized, purified single stranded DNA template suitable for sequencing with Sequenase T7 DNA polymerase (USB). This enzyme produces a very uniform band pattern and provides the choice between end-labeling of the sequencing primer and internal labeling of one dNTP in non-radioactive systems.

The sequencing reaction products were simultaneously separated and blotted onto a nylon membrane [3, 4] in the TwoStep Direct Blotter (Hoefer). This system combines automatically the steps of separation and transfer. The linear speed gradient at which the nylon membrane automatically passes under the gel provides a uniform separation and high resolution of the transferred DNA fragments. The immobilized fragments were detected using chemi-luminescent or colorimetric detection methods [5–8].

Two different methods of direct non-isotopic sequencing of PCR products will be demonstrated using the 423 bp PCR amplified genomic DNA fragment of the HMG box of the human SRY, the sex-determining region of the Y chromo-some [9]:

The forward strand will be analyzed by cyclic sequencing using the 5'-biotin-endlabeled primer AP1 [10], the reversed strand will be analyzed by solid phase sequencing using the non-labeled primer AP2 [10] and an internal labeling step with digoxigenin-16-dATP.

Both biotin and digoxigenin are high efficient non-radioactive labeling substances in manual sequencing.

The biotinylated fragments are detected via binding to streptavidin alkaline phosphatase conjugate using the chemiluminescent substrate CSPD. The light emission from the dephosphorylating CSPD results in a DNA band pattern that is captured on X-ray film.

The digoxigenated fragments are visualized in a similar way, but the streptavidin alkaline phosphatase is replaced by anti-digoxigenin alkaline phosphatase conjugate.

The quality of the band pattern is similar to that achieved through the incorporation of nucleotides labeled with ^{32}P or ^{35}S radioisotopes. Due to the short exposure times on the X-ray film, the data can be obtained in only one day.

1.4.2 Experimental Procedures

1.4.2.1 Direct Non-Isotopic Cyclic Sequencing of Double Stranded PCR Products

Materials:

(A) Sequencing Reaction

1) *PCR product*
    423 bp PCR amplified genomic DNA fragment of HMG box of SRY (5 fmol/μl). PCR amplification of genomic DNA samples (200 ng) was performed in 50 μl reactions using 0.5 μmol/l of each primer AP1 (5'-GAA-TAT-TCC-CGC-TCT-CCG-G-3') and AP2 (5'-ACA- ACC-TGT-CCA-GTT-

1.   PCR with two non-labeled primers

2.   Purification of the PCR product   (Removing primers, dNTPs)

3.   Cyclic sequencing   (5'-biotinylated primer, d/ddNTPs, Taq-polymerase)

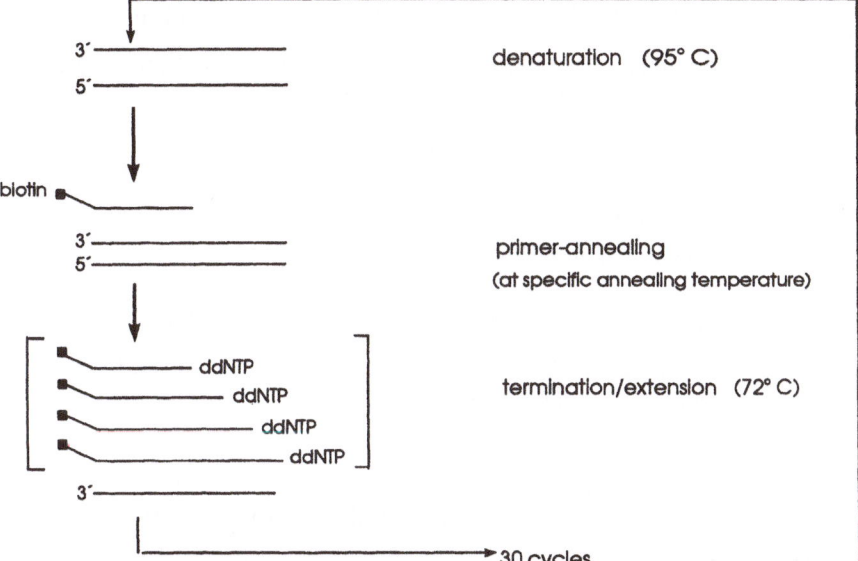

4.   Separation and parallel blotting of the 5'-biotinylated fragments
     onto nylon membrane

5.   Chemiluminescent detection

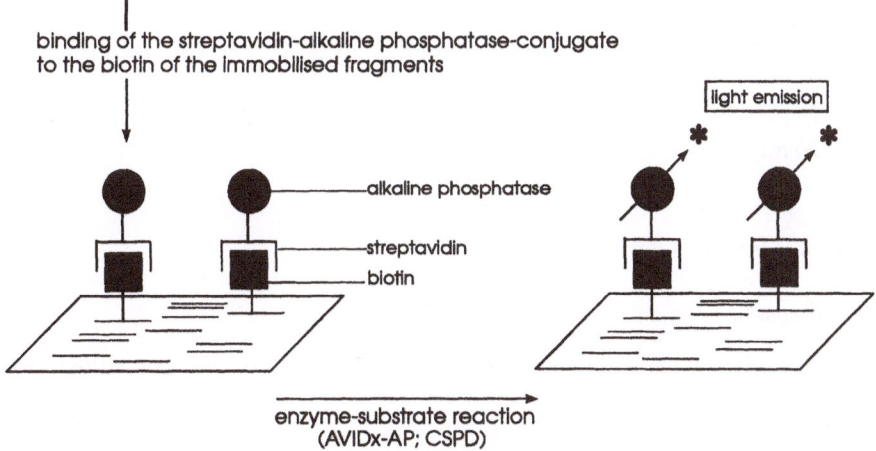

Fig. 1.4-1. Direct non-isotopic cyclic sequencing of double stranded PCR products – principle of the method

GC-3'; according to Jäger et al. [10], 3 mmol/l MgCl$_2$, 200 mmol/l each dNTP and 1.25 U of Taq polymerase.
The PCR products were purified using the QIAquick Spin PCR Purification Kit (Qiagen).

2) *Sequencing primer*
AP1: 5'-biotin-GAA-TAT-TCC-CGC-TCT-CCG-G-3', 1 pmol/µl

The following reagents [3–6] are components of the fmol DNA Sequencing System (Promega):

3) *d/ddNTP nucleotide (Deaza-) Mixes*
4) *fmol Sequencing 5x buffer*
5) *Sequencing grade Taq-polymerase*
6) *fmol Sequencing Stop Solution*

(B) Denaturing Polyacrylamide Gel Electrophoresis/Blotting

7) *1 × TBE buffer prepared from the 10 × stock solution (see appendix)*
8) *6 % Long Ranger gel (AT Biochem)*

(C) Chemiluminescent Detection of DNA Fragments

The following reagents are components of the SEQ-Light DNA Sequencing Detection Kit (Tropix).

9) *Blocking buffer:* 0.2% I-Block reagent; 0.5% SDS in 1 × PBS (0.058 mol/l Na$_2$HPO$_4$, 0.017 mol/l NaH$_2$PO$_4$ × H$_2$O, 0.068 mol/l NaCl; pH 7.4)
10) *Conjugate solution:* 1:5000 dilution of AVIDx-AP Conjugate in Blocking buffer.
11) *Wash buffer:* 0.5% SDS in 1 × PBS
12) *Assay buffer:* 0.1 M diethanolamine; 1 mmol/l MgCl$_2$ in aqua ster., pH 10
13) *CSPD detection solution:* 0.24 mmol/l CSPD in Assay buffer (200 µl in 20 ml)

Protocol 1: Direct Non-Isotopic Cyclic Sequencing

(A) Sequencing Reaction

1. For each set of sequencing reactions, label four 0.5 ml Eppendorf tubes (G,A,T,C).
   Add 2 µl of the appropriate d/ddNTP mix to each tube. Cap the tubes and store on ice until needed.
2. Mix the following reagents in an Eppendorf tube:
   - Aqua sterile 7.5 µl
   - Sequencing 5 × buffer 5.0 µl
   - PCR product 2.0 µl
   - Primer 1.5 µl (final volume 16 µl)

The TaqMan™ LS-50B PCR Detection System takes quantitative sequence detection to new levels of speed and throughput.

On top of that, specificity is higher, too. As few as 10 molecules have been detected – in PCR solution – using an energy transfer between two fluorescent dyes on a target-specific probe. To ensure maximum reliability and convenience, fluorescence is detected by the research-proven LS-50B

Luminescence Spectrometer. No gel electrophoresis, ethidium bromide, or autoradiography is required. Imagine what you can do with the TaqMan™ LS-50B PCR Detection System.

Contact your local Perkin-Elmer office for our report describing quantitation of PCR template starting copy number.

The fluorogenic probe anneals specifically to the target sequence and emits light upon cleavage during PCR.

Mix briefly by pipetting up and down.

3. Add 1.0 µl of Sequencing grade Taq polymerase to the primer/template mix. Mix briefly by pipetting up and down.
4. Add 4 µl of the enzyme/primer/template mix from steps 2. and 3. to the inside wall of each tube containing d/ddNTP mix.
5. Add 25 µl of mineral oil to each tube and briefly spin in a microcentrifuge.
6. Place the reaction tubes in the thermal cycler (e.g. TRIO Thermoblock; Biometra) that has been preheated to 95 °C and start the cycling program:
   95 °C for 5 min, then:
   95 °C for 30 s (denaturation)
   57 °C for 30 s (annealing)
   72 °C for 60 s (extension)
   30 cycles total, then
   72 °C for 5 min, then
   4 °C
7. After the thermocycling program has been completed, add 3 µl of Sequencing Stop Solution.
8. The samples may be stored at – 20 °C overnight.

(B) Denaturing Polyacrylamide Electrophoresis/Blotting

1. Pour about 1.0 liter of 1 × TBE into the lower buffer chamber.
2. Cut a piece of nylon membrane 50 cm long (Pall Biodyne A; Hoefer).

>>**Always wear gloves when handling nylon membrane**<<

3. Squirt a few milliliters of buffer under the membrane flap on the transport belt. Place one end of the nylon membrane under the flap. Advance the transport belt until the flap is located under the glass rod immediately behind the sandwich.
4. Place the gel sandwich and the aluminum plate into the blotter. Carefully inspect the interface between the bottom of the gel and the nylon membrane. If any bubbles are trapped, remove them by directing a stream of buffer at the interface using a transfer pipette.
5. Pour the buffer into the upper buffer chamber.
6. Remove the blank comb.
7. Insert the shark's tooth comb.
8. Activate the power supply for the prerun (35 W for 45 min).
9. Load the samples:
   Heat the samples at 80 °C for 2 min immediately before loading, then place them on ice. Load 2.0 µl of each reaction on the gel in the order G-A-T-C.
10. Run the gel at 30 W until the tracking dye has reached the bottom of the gel.
11. Program the Controller:
    Using MAIN MENU/EDIT/SINGLE RUN set the following program:
    Total 03.30 (h:min)
    Delay 00.00 (h:min)

Init 15.00 (cm/h)
Last 08.00 (cm/h)

Generally, the first fragments move off the gel and onto the membrane quickly; in order to allow for adequate separation of the fragments, the initial speed setting should therefore be rather high. Longer fragments move more slowly through the gel. In order to get the best resolution and still achieve adequate separation, the last speed setting should be slower than the initial speed setting.

Save the program using MAIN MENU/FILER/SAVE.

Load the program using FILER MENU/LOAD.

Activate AUTO-ADVANCE using the OPTIONS MENU in order to move the transport belt an additional 10 cm -at top speed- at the end of the run.

12. After the tracking dye has reached the bottom of the gel, start the run of the nylon membrane using MAIN MENU/START RUN.
13. Remove the membrane from the flap at the end of the run.
14. Immobilize the DNA at the membrane by UV irradiation.

(C) Chemiluminescent Detection of DNA Fragments

The following steps will be performed at room temperature in a hybridization oven (MWG).

1. Place the membrane in the tube (GATC).
2. Add 100 ml of Blocking buffer and incubate for 15 min.
3. Incubate for 20 min in 100 ml of Conjugate solution.
4. Wash 1×5 min in 100 ml Blocking buffer.
5. Wash 3×10 min in 100 ml Wash buffer.
6. Wash 2×1 min in 100 ml Assay buffer.
7. Incubate for 5 min in CSPD Detection solution.
8. Remove the membrane and seal the membrane in plastic wrap. Do not allow the membrane to dry.
9. Expose for 30 min at room temperature to BIOMAX MR film (Kodak).

1.4.2.2 Direct Non-Isotopic Solid-Phase Sequencing of Single Stranded PCR Products

Materials:

(A) Separation of Single-Stranded DNA

1) PCR product
423 bp PCR amplified genomic DNA fragment of HMG box of SRY. PCR amplification of genomic DNA samples (500 ng) was performed in 50 µl reactions using 0.5 µmol/l of each primer AP1 (5′-biotin-GAA-TAT-TCC-CGC-TCT-CCG-G-3′) and AP2 (5′-ACA-ACC-TGT-CCA-GTT-GC-3′; according to Jäger et al. [10], 3 mmol/l MgCl$_2$, 200 mmol/l each dNTP and 1.25 U of Taq-polymerase.

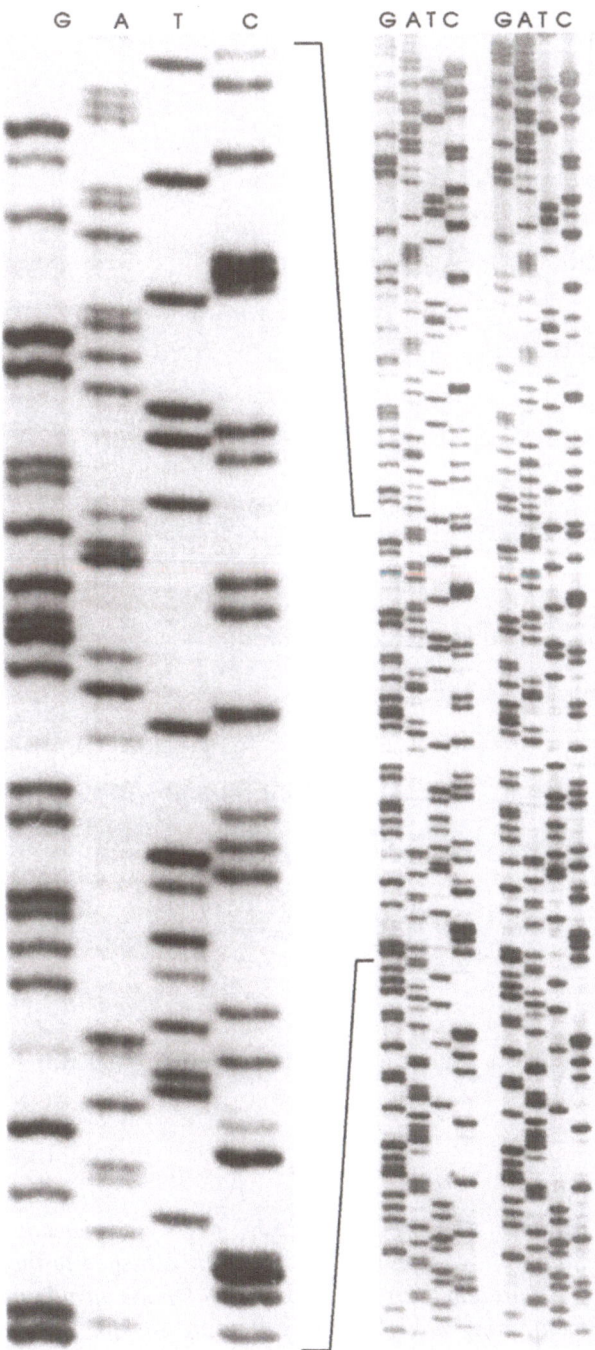

Fig. 1.4-2. Direct non-isotopic sequencing of double stranded PCR products using a 5′-biotinylated primer. DNA sequences generated with the fMol DNA Sequencing System (Promega) as detailed in the protocol 1

1. PCR with one nonlabeled and one 5'-biotinylated primer
2. Preparation of single-stranded DNA

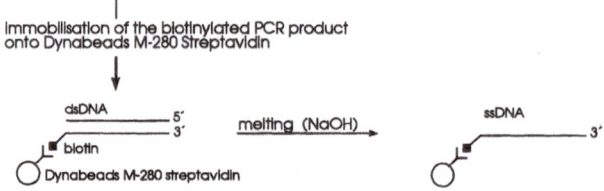

3. Solid-phase sequencing (ssDNA, DIG-16-dATP, Sequenase 2.0, d/ddNTPs)

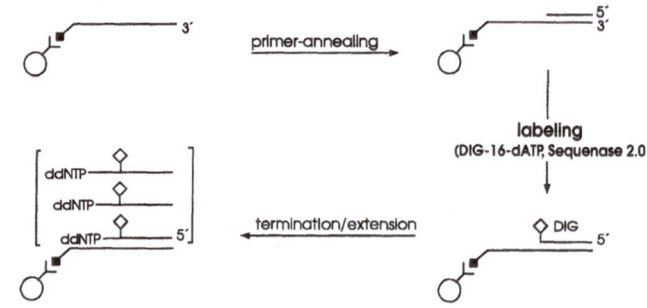

4. Separation and parallel blotting of the DIG-labeled fragments onto nylon membrane
5. Chemiluminescent detection

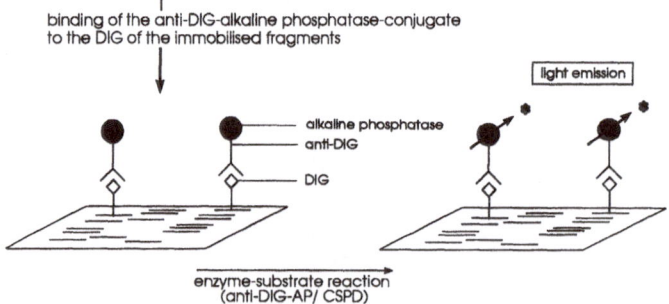

Fig. 1.4-3. Direct non-isotopic solid-phase sequencing of single stranded PCR products – principle of the method

2) *Dynabeads M-280 Streptavidin*
   Dynabeads M-280 are uniform, superparamagnetic, polystyrene beads. Dynabeads M- 280 Streptavidin have streptavidin covalently attached to the bead surface. Streptavidin is a protein (MW of approx. 66 000) made up of four identical subunits, each containing a high affinity binding site for biotin $(K_D = 10^{15} \, mol/l)$.
3) *Binding & Washing buffer (B & W, Dynal): 10 mmol/l Tris-HCl, pH 7.5; 1 mmol/l EDTA, 2.0 mol/l NaCl*
4) *NaOH-solution: 0.1 mol/l NaOH, 0.04 mmol/l EDTA*
5) *TE-buffer: 10 mmol/l Tris-HCl, pH 7.5; 1 mmol/l EDTA*

(B) Solid-Phase Sequencing

6) *Sequencing Primer*
    AP-2: 5'-ACA-ACC-TGT-CCA-GTT-GC-3', 10 pmol/µl
7) *Labeling substance: DIG-16-dATP (Boehringer, Mannheim)*
    17 µmol/l solution of digoxigenin-16-dATP in aqua sterile.
    Final concentration in the labeling mixture: 1 µmol/l
    DIG-16-dATP can replace dATP as a substrate for T7 DNA polymerase in
    DNA sequencing according to Sanger et al. [2].

See Note at the end of the chapter.

The following reagents are components of the Sequenase Version 2.0 DNA
Sequencing Kit (United States Biochemicals):

8) *Sequenase reaction buffer, 5×concentrated*
9) *Dithiotreitol, 0.1 mol/l*
10) *Mn buffer*
11) *Enzyme dilution buffer*
12) *G/A/T/C-termination/extending mixes: 1 Vol. Extending Mix, 4 Vol*
    Termination Mix
13) *Sequenase Version 2.0 T7 DNA polymerase, 13 U/µl*
14) *Stop solution*

(C) Denaturing Polyacrylamide Electrophoresis/Blotting

15) *1×TBE buffer prepared from the 10 ×stock solution (see appendix)*
16) *6% Long Ranger gel (AT Biochem)*

(D) Chemiluminescent Detection of DNA Fragments

15) *1×PBS: 0.058 mol/l Na_2HPO_4, 0.017 mol/l $NaH_2PO_4 × H_2O$, 0.068 mol/l NaCl*
    in aqua bidest; pH 7.4
16) *Blocking buffer: 0.2% I-Block reagent; 0.1% Tween 20 in 1×PBS*
17) *Conjugate solution:*
    1:5000 dilution of Anti-DIG-AP conjugate (polyclonal sheep anti-
    digoxigenin Fab-fragment, conjugated with alkaline phosphatase;
    Boehringer, Mannheim) in 1 volume of 1×PBS and 1 volume Blocking
    buffer.
    To avoid unspecific spots at the blot, spin the antibody for 3 min in a micro-
    centrifuge before use, then dilute with 1 vol. 1 PBS, filtrate the solution with
    a syringe through a 0.45 µm microfilter and add the Blocking buffer.
18) *Wash buffer: 0.3% Tween 20 in 1×PBS*
19) *Assay buffer: 0.1 mmol/l diethanolamine; 1 mmol/l $MgCl_2$ in aqua ster.,*
    pH 10
20) *CSPD Detection solution: 0.24 mmol/l CSPD in Assay buffer (200 µl in 20 ml)*

Protocol 2: Direct Non-Isotopic Solid Phase Sequencing

(A) Preparation of Single Stranded DNA

1. Place the tube containing 40 µl of Dynabeads in the appropriate magnetic particle concentrator (MPC).
   Remove the supernatant with a pipette while keeping the tube in the magnet.
2. Resuspend the Dynabeads in 40 µl of B & W buffer and mix gently by pipetting up and down. Remove the supernatant with a pipette while keeping the tube in the MPC.
3. Repeat 2.
4. Resuspend the beads in 40 µl B & W buffer.
5. To the 40 µl of prewashed beads add 40 µl of PCR product. Mix by pipetting up and down.
6. Place the tube in the thermocycler (e.g. TRIO Thermoblock, Biometra) that has been preheated to 37 °C.
7. Incubate for 30 min keeping the beads suspended by gently tipping the tube.
8. Remove the supernatant using the MPC.
9. Wash the Dynabeads/PCR-product twice with 2×40 µl B & W buffer using the MPC.
10. Remove the supernatant using the MPC.
11. Add 50 µl of the NaOH solution. Mix by pipetting up and down.
12. Incubate at room temperature for 5 min.
13. Remove the supernatant using the MPC.
14. Wash the beads (with the immobilized biotinylated strand) once with 50 µl NaOH solution, once with 50 µl B & W buffer and once with 50 µl TE buffer.
15. Remove the supernatant using the MPC.
16. Resuspend the beads in 8 µl aqua sterile.

(B) Solid-Phase Sequencing

Primer annealing:

1. To the tube containing the immobilized ssDNA (10 µl) add: 2 µl Reaction buffer and 1 µl Sequencing primer (10 pmol).
   Mix by pipetting up and down.
2. Heat the sample for 5 min at 65 °C in the thermocycler, slowly cool down to room temperature over a period of 30 min.

Labeling reaction:

3. While the mixture is cooling down dilute the Sequenase enzyme to 2.6 U/µl: Mix 2.0 µl Enzyme dilution buffer and 0.5 µl Sequenase enzyme in a fresh tube.
   Keep the diluted enzyme on ice.
4. After the template-primer-reaction mix has cooled down add (on ice):
   - 1 µl dithiotreitol
   - 1 µl Mn buffer

- 1 μl DIG-dATP labeling solution and
- 1 μl diluted Sequenase enzyme (2.6 U)

5. Incubate for 10 min at 37 °C.
6. After the labeling reaction has finished add another 1 μl aliquot of the diluted Sequenase enzyme (2.6 U).

Termination/extension reaction:

7. While the labeling mixture is incubated at 37 °C, label four fresh 0.5 ml tubes (G, A, T, C).
   Add 2.5 μl of the appropriate termination/extending mixes to each tube and bring up to 37 °C.
8. Add 3.5 μl of the labeling mixture to each of the four termination mixes.
9. Incubate for 5 min at 37 °C.
10. Stop the reaction by placing the tubes on ice. The reaction products might be stored at 4 °C until the next day.

Preparation of the samples for electrophoresis:

11. Mix the samples by pipetting up and down.
12. Place the samples into MPC and remove the supernatant.
13. Add 4 μl Stop solution. Mix gently.
14. Heat the samples for 2 min to 80 °C and chill on ice prior to applying to the gel.

(C) Denaturing Polyacrylamide Electrophoresis/Blotting

The procedure corresponds to the method described in protocol 1 part (B) p. 43-4

(D) Chemiluminescent Detection of DNA Fragments

The following steps will be performed at room temperature in a hybridization oven (MWG).

1. Place the membrane in the tube.

>>Always wear gloves when handling nylon membrane<<

2. Add 100 ml 1 × PBS and incubate for 3 min.
3. Incubate for 20 min in 100 ml Blocking buffer.
4. Incubate for 20 min in 100 ml of Conjugate solution.
5. Wash 3 × 10 minutes with Wash buffer.
6. Equilibrate the membrane for 2 × 1 min with 100 ml Assay buffer.
7. Incubate for 5 min in 20 ml CSPD Detection solution.
8. Expose for 30 min at room temperature to BIOMAX MR film (Kodak)

Note: The digoxigenin residue will slightly change the migration of the DNA fragments through the sequencing gel. Consequently, identical numbers of labeled nucleotides must be incorporated into each DNA fragment to provide for a uniform, single banding pattern and to avoid the appearance of multiple background bands.

The best results will be achieved by incorporating the digoxigenin residue only once into each strand by designing a labeling mixture containing the DIG-16-dATP and missing the unmodified dATP and at least one of the other three nucleotides.

Example 1:

primer
5′ |—————|GCA^{DIG}
3′ |—————CGTA.......
template

If the labeling mixture contains dGTP, dCTP and DIG-16-dATP but no dTTP, the primer is extended by three bases, the first two being unmodified and the third digoxigenated.

Example 2:

primer
5′ |—————|GA^{DIG}
3′ |—————CTGG.......
template

In the presence of only dGTP and DIG-16-dATP in the labeling mixture the primer is extended by two bases, the first being unmodified and the second digoxigenated.

Example 3:

Sequencing of PCR amplified SRY-HMG box by the primer AP 2 (present case)

primer
5′ |AC-AAC-CTG-TTG-TCC-AGT-TGC|-A^{DIG}
3′ -TG-TTG-GAC-AAC-AGG-TCA-ACG-TGA-AGC.......
template

The first nucleotide to be linked with the 3′end of the sequencing primer is dATP. Thus the labeling mixture contains exclusively DIG-16-dATP. During the labeling step the primer is extended by only one base, the digoxigenated dATP.

In either case the labeling reaction is concluded with the incorporation of one, and only one labeled nucleotide into the newly synthesized DNA strand.

The remaining unincorporated DIG-dATP will be diluted out during the following termination/extension step due to the excess of unlabelled dATP in the reaction mixture.

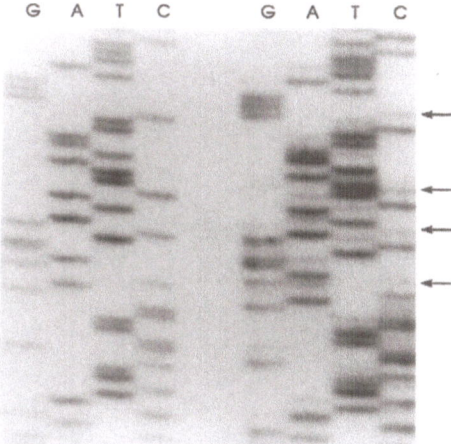

Fig. 1.4-4. Direct non-isotopic sequencing using an internal labeling method with DIG-16-dATP.
Left: DNA sequences generated with the Sequenase 2.0 (USB) as described in the protocol 2.
Right: DNA sequencing pattern using the same protocol but a labeling mixture containing the following substances: 17 µmol/l DIG-16-dATP; 1 µmol/l dGTP; 1 µmol/l dCTP and 1 µmol/ldTTP. All bands are accompanied by slightly weaker artifact bands (←) due to the slower migration of the population of multiple labeled DNA fragments

References

1. Saiki AK, Gelfand DH, Stoffel S, Scharf SJ, Higuchi R, Horn GT, Mullis KB, Erlich HA (1988) Primer-directed enzymatic amplification of DNA with a thermostable DNA polymerase. Science; 239:487–491
2. Sanger F, Nicklen S, Coulson AR (1977) DNA sequencing with chain-terminating inhibitors. Proc Natl Acad Sci; 74:5463–5467
3. Pohl FM, Beck S (1987) Direct transfer electrophoresis used for DNA sequencing. Methods Enzymol; 155:250
4. Richterich P, Heller C, Wurst H, Pohl FM (1989) DNA sequencing with direct blotting electrophoresis and colorimetric detection. BioTechniques; 7:52–59.
5. Bronstein I, Edwards B, Voyta JC (1989) 1,2-Dioxetanes: Novel chemiluminescent enzyme substrates. Applications to immunoassays. J Biolumin Chemilumin; 4:99–111.
6. Höltke HJ, Sanger G, Kessler C, Schmitz G (1992) Sensitive chemiluminescent detection of digoxigenin-labeled nucleic acids: A fast simple protocol and 1st applications. Biotechniques; 12:104–113.
7. Martin C, Bresnick L, Jou RR, Voyta JC, Bronstein I (1991) Improved chemiluminescent DNA sequencing. BioTechniques; 11:110–113.
8. Tizard R, Cate RL, Ramachandran KL, Wysk M, Voyta JC, Murphy OJ, Bronstein I (1990) Imaging of DNA sequences with chemiluminescence. Proc Natl Acad Sci USA; 87:4514–4518.
9. Sinclair AH, Berta P, Palmer MS, Hawkins JR, Griffiths BL, Smith MJ, Foster JW, Frischauf A-M, Lovell-Badge R, Goodfellow PN (1990) A gene from the human sex-determining region encodes a protein with homology to a conserved DNA-binding motif. Nature; 346:240–244
10. Jäger RJ, Anvret M, Hall K, Scherer G (1990) A human XY female with a frame shift mutation in the candidate testis-determining gene SRY. Nature; 348:452–454

Part 2
Conventional Techniques for mRNA Analysis

Chapter 2.1
Isolation of mRNA

D. Lassner

2.1.1 Theoretical Background

Gene expression and gene regulation are the basis of all cell and tissue functions and are bound up to transcriptional and translational processes.

The conversion of biological information from genomic sequences (double-stranded DNA) into a functional polypeptide through expression of heterogeneous nuclear RNA in the nucleus, splicing of mRNA sequences, posttranscriptional and posttranslational modifications is a field of extensive research in molecular biology and clinical diagnostics.

The typical mammalian cell contains approximately $5 \times 10^{-5} \mu g$ RNA [1]. There are five different forms of RNA with various functions in gene expression. All species of cellular RNA are transcriptional products of one of the three nuclear RNA polymerases (see Table 2.1-1).

During transcription, RNA polymerase II produces RNA molecules whose sequence correlates precisely with DNA from which it is derived [2]. All these transcribed RNA precursors contain introns and are called heterogeneous nuclear RNA (hn RNA). During formation of mRNA, sequence introns are removed by splicing and a poly A^+ tail is added to the 3′-ends. The length of the

Table 2.1-1. Eukaryotic RNA polymerases, transcriptional products, mean functions and relative contribution of cellular RNA

Enzyme	Transcriptional products	Function	percentage of cellular RNA (%)
RNA polymerase I	ribosomal RNA (rRNA, 28 S, 18 S, 5.8 S)	constitution of ribosomes	70–80
RNA polymerase II	heterogeneous nuclear RNA (hn RNA)	precursor of mRNA	20–40
RNA polymerase III	transfer RNA (tRNA)	translation of mRNA to protein sequence	10
	ribosomal RNA (rRNA)	constitution of ribosomes, intron removal	
	small nuclear RNA (snRNA)	and splicing of hn RNA	

poly A segment is quite consistent: 200–250 nucleotides in vertebrates, about 100 nucleotides in yeast [3].

Heterogeneous nuclear RNA is synthesized in the nucleus. The relative contribution of its spliced form, namely mRNA, is only 2–5% of the total cellular RNA. This indicates that about 75% of all hnRNA is degraded in the nucleus.

Almost all eukaryotic mRNAs are monocistronic. This means they encode a single polypeptide chain. In prokaryotes, many mRNA are polycistronic, i.e. they encode multiple polypeptide chains [3].

Diversity of cellular biochemistry corresponds to the diversity of cellular RNA. Thus, mRNA is of particular interest to molecular biologists and in many fields of clinical diagnostics [2].

2.1.2 Precautions in RNA Isolation

With extensive research in the field of molecular biology, the isolation of pure RNA from a large number of biological samples has become a critical issue. The crucial aspect in any RNA isolation procedure is protection of the sample from contamination with ribonucleases. RNases are very stable enzymes and generally do not require any cofactors for their function. Therefore a small amount of RNase in a RNA sample is a real problem. Under experimental conditions, the hands of the researcher are the major source of contamination with RNases. Therefore:

>> **Always wear gloves when handling RNA.<<**
>>**Do not hesitate to change gloves frequently.<<**

Autoclaving will not fully inactivate RNases. Water or salt solutions used in RNA preparation should be treated with 0.1% diethylpyrocarbonate (DEPC) for at least 1 hour. Then autoclave treated solutions to remove any traces of DEPC. This chemical inhibits ribonucleases irreversibly. Tris buffers cannot be treated with DEPC as Tris react with it. Therefore Tris should be dissolved in DEPC-treated- and autoclaved water to get an RNase-free solution. The solution should be autoclaved again after addition of Tris.

>>**When using DEPC, wear gloves and use a fume hood because DEPC is a strong carcinogen!<<**

Sterile, disposable plasticware is essentially free of RNase activity. However, autoclave all plasticware before first use and dry it at 110 °C for 4 h to remove excess moisture. Glassware is heated to 210 °C for 4 h to destroy all ribonuclease activity. These measures are sufficient to obtain RNase-free conditions. Many laboratory manuals also recommend rinsing all plasticware with chloroform before autoclaving.

All chemicals should be of the highest purity.

2.1.3 Methods of mRNA Isolation

Guanidinium thiocyanate and chloride are among the most effective protein denaturants. They are both strong inhibitors of ribonucleases and guanidinium extraction has become the method of choice for RNA purification.

Most common and consistently successful methods for isolation of pure, intact total RNA are modifications of the original guanidinium thiocyanate method of Chirgwin et al. [4].

This method has been used to isolate undegraded RNA from ribonuclease-rich tissue e. g. pancreas. The protocol combines the disruption of tissue or cells in high concentration of guanidinium thiocyanate and a subsequent ultra-centrifugation of this lysate through a CsCl cushion. The RNA forms a pellet at the bottom of centrifugation tube, while proteins and DNA remain in or above the CsCl cushion [5].

A modified method combining guanidinium thiocyanate and phenol-chloroform extraction does not require an ultracentrifugation step [6]. This is often the method of choice when multiple RNA extractions are performed. The main component of this procedure is solution D (25 mmol/l sodium citrate, pH 7.0, 4 mol/l guanidinium thiocyanate, 0.5% sarcosyl, 0.1 mol/l 2-mercaptoethanol). The extraction step is performed by using a mixture of solution D, phenol, 0.2 mol/l sodium acetate; pH 4.0, and chloroform. Two phases are obtained after incubation on ice and centrifugation at 10000 g for 20 min. The upper aqueous phase contains the RNA, whereas proteins and DNA are in the lower phenol phase and the inter-phase. After addition of an equal volume of isopropanol to the separated upper aqueous phase, the RNA precipitates at − 20 °C for at least 1 h. Sedimentation at 10000 g for 20 min was again performed and the resulting RNA pellet was resolved in solution D and reprecipitated with one volume of isopropanol. After centrifugation RNA pellet was resuspended in 75% ethanol, sedimented, vacuum dried, and dissolved in DEPC-treated water or 0.5% SDS. Ribonucleases are inhibited during this protocol by the presence of 4 M guanidinium thiocyanate.

Both isolation protocols described here provide high yield and purity of undegraded RNA preparations. An improved protocol for a rapid single-step method of RNA isolation is demonstrated below. Here a ready-to-use RNAzol B-solution which combine all components and properties of solution D and phenol in one solution, resulting in a more convenient protocol.

Purified total RNA could be used for many downstream applications (i. e. poly A+ RNA selection by oligo (dT) chromatography, Northern blot analysis (see chapter 2.6) and cDNA synthesis (chapter 2.2)).

For some applications, it is more convenient to use mRNA instead of total cellular RNA because of increased sensitivity, especially in Northern blot analysis (see chapter 2.6).

The mRNA can be isolated from total RNA by oligo (dT) chromatography. There are existing protocols to isolate mRNA directly from cell lysates. One of most convincing and reliable methods of mRNA isolation is the magnetic separation method with oligo (dT) bound on the surface of paramagnetic beads. A protocol for isolation of polyA+ RNA directly from cell lysate is described below.

2.1.4 Experimental Procedures

All protocols mentioned below were established for isolation of RNA from cultured cells. Disruption of tissues is performed by mincing the freshly isolated sample (up to 0.1 g) on ice and subsequent homogenization with a glass/Teflon homogenizer with 1 ml of corresponding lysis solution (RNAzolB or lysis buffer) at room temperature. All the following steps were adapted to a scaled-up protocol according to the volume of lysed cells/tissues. For tissues difficult to homogenize, it is recommend to put the isolated sample first in liquid nitrogen, grind the frozen material and transfer the frozen powder to a glass/Teflon homogenizer containing the lysis solution (RNAzol B or lysis buffer). Thereafter the procedure mentioned above is followed.

2.1.4.1 Isolation of Total-RNA by RNAzol B

RNAzol B is a ready-to-use solution for rapid isolation of total-RNA from cells and tissues. It contains an irritant (guanidinium thiocyanate) and a poison (phenol). Handle RNAzol B with gloves. There are many comparable chemicals (e. g. TRISOLV, TRIzol) for RNA isolation with slightly improved properties (e. g. isolation of RNA and DNA in one extraction step).

  The following protocol is the method of choice for rapid RNA isolation. It is easily scaled up or down depending on sample size.

Materials:

1) *RNAzol B (AGS)*
2) *Chloroform (– 20 °C)*
3) *Isopropanol (– 20 °C)*
4) *Ethanol 75 (v/v) in H_2O,(– 20 °C)*
5) *DEPC-H_2O*
6) *Eppendorf reaction tubes (Eppendorf)*
7) *Eurotips of different size (Eppendorf)*
8) *Varipettes 4810 (Eppendorf)*
9) *Centrifuge 5415C (Eppendorf)*
10) *Thermomixer 5436 (Eppendorf)*
11) *Heto vacuum concentrator (Heto)*
12) *Refrigerator (– 20 °C)*
13) *Iso-Rack (0 °C) (Eppendorf)*

Protocol 1: Isolation of Total Cellular RNA by RNAzol B Solution

1. Homogenization
    Lyse cells by addition of 0.2 ml RNAzol B per 10^6 cells. Solubilize RNA by passing the lysate through the pipette or by vortexing.

Neues quantitatives PCR[1] System
– PCR-Light® von TROPIX –

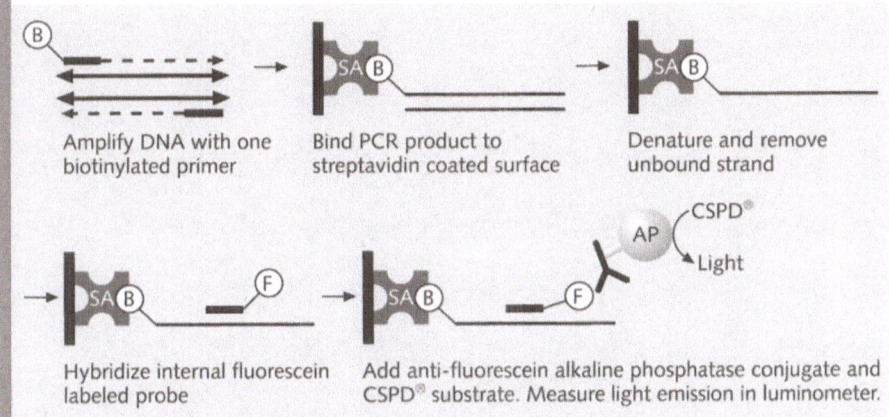

Amplify DNA with one biotinylated primer

Bind PCR product to streptavidin coated surface

Denature and remove unbound strand

Hybridize internal fluorescein labeled probe

Add anti-fluorescein alkaline phosphatase conjugate and CSPD® substrate. Measure light emission in luminometer.

Das neue PCR-Light® System von TROPIX bietet eine schnelle Chemilumineszenz-Methode für die Quantifizierung von DNA oder RNA, die mittels „Polymerase Chain Reaction" (PCR[1]) amplifiziert wurden.

- Ultrasensitive nicht-radioaktive Chemilumineszenz Methode
- Akkurate Quantifizierung von PCR-Produkten über einen weiten Konzentrationsbereich von Ausgangssequenzen
- Durchführbar auf Mikrotiterplatten
- Quantifizierung in weniger als 2 Stunden für 96 Proben möglich
- Keine Detektion von Primer Dimeren

Wählen Sie das neue PCR-Light® System für die Quantifizierung Ihrer PCR-Produkte. Das PCR-Light® System erhalten Sie in Deutschland exclusiv von SERVA Heidelberg.

SERVA

SERVA FEINBIOCHEMICA GmbH & Co. KG
A BOEHRINGER INGELHEIM COMPANY
P. O. B. 10 52 60 · D-69042 HEIDELBERG
CARL-BENZ-STR. 7 · D-69115 HEIDELBERG
TEL.: 0 62 21 / 5 02 - 0 · FAX: 50 21 13

[1] PCR ist durch ein Patent der Firma Hoffmann La-Roche geschützt

2. RNA extraction
   Add 1/10 volume ice-cold chloroform to homogenate, cover the sample tightly and vortex vigorously for 15 s. Place the tube in an Iso-Rack (0 °C) or on ice for 5 min. Centrifuge suspension at 14000 rpm for 15 min. After addition of chloroform and centrifugation, the homogenate forms two phases: the lower blue phenol-chloroform phase and the colorless upper aqueous phase. RNA remains exclusively in the upper aqueous phase whereas DNA and proteins will be in the interphase and organic phase, respectively. One volume of the aqueous phase corresponds to about 50% of the initial volume of RNAzol B plus one volume of sedimented cells.

Note: The extraction step can also be performed with a swing out rotor at $5000 \times g$ for 15 min.

3. RNA precipitation
   Transfer the aqueous phase to a fresh Eppendorf tube, add 1 volume of isopropanol (–20 °C) and mix the solution by inverting the tube. Store the samples for at least 30 min at – 20 °C. Centrifuge the suspension at 14000 rpm for 15 min. The precipitated RNA will form a white-yellow pellet at the bottom of the tube (often not visible to the naked eye).

4. RNA wash
   Remove the supernatant immediately and add 500 µl 75% Ethanol (– 20 °C) to wash the pellet. Vortex briefly and centrifuge for 8 min at 14000 rpm. After centrifugation, remove ethanol by inverting the tube and placing it on a paper towel on the bench top. Dry the pellet briefly under vacuum for 10–15 min.

5. Redissolving of RNA
   Dissolve dried RNA pellet in 50–200 µl DEPC-H_2O or suitable buffer at 50 °C in a thermomixer. The volume of added DEPC-water depends on size of visible RNA pellet.
   Approximately 25 µg of total cellular RNA should be recovered by RNA isolation with RNAzol B from 10^6 cells.

Trouble Shooting

The most critical step is the vigorous mixing of homogenate after addition of chloroform. From the following centrifugation, a colorless upper aqueous phase and a blue-brown lower phenol phase must result. Repeat the vigorous mixing on a vortexer and centrifuge again if the phases are located inversely.

In general we recommend using RNAzol B added copiously to be sure that all cell material is fully disrupted. The yield of total RNA is increased by longer precipitation time (2 h, overnight) and lower temperature (–70 °C) during precipitation.

Using polyethylene reaction tubes for RNA isolation is recommended to avoid breakage during phenol-chloroform treatment, especially during the centrifugation step.

All RNA preparations should be examined by denaturing agarose gel electrophoresis (see chapter 2.6). Total RNA is undegraded if two clear bands of ribosomal RNA (28S and 18S) are observed with the 28S band being about twice as intense as the 18S band.

2.1.4.2 mRNA Purification by Dynabeads Oligo (dT)$_{25}$

Several years ago DYNAL AS (Oslo, Norway) developed uniform, paramagnetic monodisperse polymer particles. Today these particles are known as DYNA-BEADS and have proved themselves to be useful in solving many practical problems during the separation and isolation of molecules or cells. For the various applications, the magnetic beads can be loaded with antibodies, steptavidin or oligonucleotides covalently attached to the surface. It is therefore possible to bind specific antigens or nucleic acid sequences to the beads following separation of target molecules or cells using a magnetic particle concentrator (MPC).

Dynabeads Oligo(dT)$_{25}$ are loaded with chains of deoxythymidylate, 25 nucleotides in length, covalently bound to the surface of the beads. This product is designed for rapid isolation of pure, intact polyadenylated mRNA (polyA$^+$ RNA) from eukaryotic cells. The method relies on base pairing between oligo (dT) residue coupled to the beads and the universal polyA$^+$ tail of messenger RNA.

An important feature of this beads is the possibility to prepare mRNA directly from lysates of cells, solid tissue and plants. The kit is flexible and the protocol may be scaled up or down to specific sample requirements. The purified mRNA is well suited for Northern blot analysis (see chapter 2.6)

All the required reagents are provided with the "Dynabeads mRNA DIRECT Kit" of DYNAL A.S., Oslo, Norway (excluding PBS, RNase inhibitor, glycogen and K-acetate). The essential component of this kit, paramagnetic Dynabeads Oligo (dT), can be ordered separately .

A similar product distributed by Promega is called "PolyATtract mRNA Isolation System".

This protocol exploits the ability of the dodecylsulphate (LiDS) to inhibit RNase activity.

Materials:

1) *Lysis/Binding Buffer: (100 mmol/l Tris/HCl; pH 8.0, 500 mmol/l LiCl, 10 mmol/l EDTA; pH 8.0, 1 % LiDS, 5 mmol/l dithiothreitol)*
2) *RNase Inhibitor (20 U/μl , AGS Heidelberg)*
3) *Dynabeads Oligo (dT)$_{25}$ Beads (5 mg beads/ml PBS , 0.02 % NaN$_3$)*
4) *1 × Washing Buffer with LiDS (10 mmol/l Tris/HCl; pH 8.0, 0.15 mol/l LiCl, 1 mmol/l EDTA, 0.1 % LiDS)*
5) *1 × Washing Buffer (10 mmol/l Tris/HCl; pH 8.0, 0.15 mol/l LiCl, 1 mmol/l EDTA)*

6) *Elution Buffer (2 mmol/l EDTA; pH 8.0)*
7) *Glycogen solution (20 mg/ml, Boehringer Mannheim)*
8) *K-acetate-solution (3 mol/l)*
9) *DEPC-H_2O*
10) *Ethanol (100%, – 20 °C)*
11) *Phosphate-buffered saline (PBS): see appendix*
12) *Eppendorf reaction tubes (Eppendorf)*
13) *Eurotips of different size (Eppendorf)*
14) *Varipettes 4810 (Eppendorf)*
15) *Centrifuge 5415C (Eppendorf)*
16) *Thermomixer 5436 (Eppendorf)*
17) *Heto Vac (Heto)*
18) *Refrigerator (– 20 °C)*
19) *Magnetic particle concentrator (MPC-E, Dynal)*
20) *2 ml syringe, fitted with a 21 gauge needle*

Protocol 2: mRNA Purification by Dynabeads Oligo (dT)$_{25}$

(A) Preparation of lysate from cells

1. Wash the cell suspension with PBS prior to preparing a cell pellet by centrifugation. The cell pellet can be used immediately, or frozen in liquid nitrogen.
2. Resuspend sedimented cells (up to 4×10^6) with 1 ml Lysis/Binding buffer, add 1 µl (20 U) RNase inhibitor. Pass the solution repeatedly through a pipette tip to obtain a complete lysis.
3. Reduce the viscosity by a DNA-shear step. The lysate is pressed three times through a 21 gauge needle by a 2 ml syringe (use force). The reduced viscosity is visible. Repeated shearing causes foaming of the lysate. This should not effect the mRNA yield. Centrifuge the solution for 30 s to reduce the foam.

(B) Conditioning of Dynabeads Oligo (dT)$_{25}$

4. Place a tube with 0.25 ml Dynabeads Oligo (dT)$_{25}$ into MPC-E for 30 s and remove the supernatant. Wash the beads in 200 µl Lysis/Binding buffer. Again remove the supernatant using the MPC-E.
5. Transfer the Eppendorf tube with the prewashed Dynabeads to another rack (outside the MPC-E).

(C) Direct mRNA Isolation from Crude Lysate

6. Mix the cell lysate with Dynabeads and allow to hybridize for 5 min at room temperature.
7. Use the MPC-E to collect the beads on one side of tube for 2 min and remove the supernatant.

8. Wash the beads twice with 1 ml washing buffer with LiDS and three times with 0.5 ml washing buffer. Remove the last wash solution as completely as possible.
9. Add 20 μl 1 × Elution buffer, incubate at 65 °C for 2 min (thermomixer) and separate beads in MPC-E immediately. Transfer supernatant (Caution! Contains mRNA) to a new RNAse free Eppendorf tube.

Note: This purified mRNA can be used directly for Northern blotting or RT-PCR. If desired precipitate RNA performing steps 10 to 12.

10. Precipitate RNA after addition of 290 μl elution buffer, 10 μl glycogen solution, 30 μl K-acetate solution and 700 μl Ethanol for 30 min at – 20 °C as a minimum.
11. Centrifuge mRNA for 15 min at 14000 rpm. Remove supernatant by inverting the tube and placing on a paper towel on the bench top.
12. Dry the mRNA pellet briefly under vacuum for 5 min. Dissolve dried pellet in DEPC- H_2O or suitable buffers at 50 °C in a thermomixer.
    This protocol is suitable for isolating between 1 and 3 μg of poly A+ RNA.

Trouble Shooting

Examination of mRNA samples on denaturing agarose gel (see chapter 2.6) will exhibit two weak bands (28S and 18S rRNA). No mRNA preparation is free of ribosomal RNA, but rRNA does not influence further applications.

The added RNase inhibitor protects this preparation only in the first step of this protocol.

It is necessary to perform the RNA isolation as careful as possible to prevent contamination with ribonucleases.

2.1.5 Quantitation of Purified mRNA

The concentration of isolated mRNA in the final eluate can be determined by spectrophotometry. Estimation of the amount of mRNA is performed by measuring absorbance at 260 nm and 280 nm.

The absorbance at 260 nm corresponds to the amount of nucleic acids (DNA, RNA) and proteins in the sample. Absorbance at 280 nm only determines the concentration of proteins. The A_{260}/A_{280} ratio should be between 1.8 to 2.0 when the RNA preparation is free of proteins. Nucleic acid preparation is sufficiently free of proteins when the A_{260}/A_{280} ratio has a value of about 1.6.

If the ratio is lower than 1.6 remove contaminating proteins by phenol-chloroform extraction as described above (chapter 1.3).

Pipette dilutions of RNA in quartz cuvettes and measure absorbance at 260 nm and 280 nm of each dilution by photometry and calculate the concentration of RNA using the following equation:

Concentration of RNA (ng/µl) = A_{260} * dilution factor * extinction coefficient
Extinction coefficients of nucleic acids are (width of cuvette is 10 mm):

ds-DNA   50.0
ss-DNA, RNA   40.0 (more precisely 44.19 for RNA) [2]
oligonucleotides 25.0

A_{260} of 1.00 (1 O.D., d = 10 mm) correspond to a concentration of 40 µg RNA per ml examined solution.

The dilution factor is usually between 50 to 400 depending on the size of quartz cuvettes used. For minimizing contamination, it is preferable to measure the concentration of nucleic acids by spectrophotometry and to discard the diluted sample. If reusing this sample is necessary, clean the cuvettes by soaking with chromic acid (or concentrated HCl:methanol (1:1)) and then rinse thoroughly with sterile DEPC-treated water to purge of RNase activity.

2.1.6 Storage of Purified RNA

For long-term storage of isolated RNA, a washed and dried pellet is covered with ethanol (100%) and stored at −20 °C, or in aqueous solution at −70 °C or below for up to one year without appreciable deterioration [3]. Avoid repeated thawing and freezing.

For daily use, isolated RNA is aliquoted and can be stored in aqueous solution at −20 °C without significant loss of stability. To ensure intactness and longer storage time of RNA preparations, add a carrier RNA (e.g. tRNA of *E. coli*). For mRNA molecules, the first sign of aging and instability is a shortage of or a lost poly A^+ tail.

References

1. Sambrook J, Fritsch EF, Maniatis T (1989) Molecular Cloning: A laboratory manual. Cold Spring Harbor University Press, Cold Spring Harbor
2. Farell RF (ed.) (1993) RNA Methodologies: A Laboratory Guide for Isolation and Characterization, Academic Press Inc., New York
3. Darnell JE (ed.), Lodish H, Baltimore D (1990) Molecular Cell Biology. Scientific American Books Inc., New York
4. Chirgwin JM, Przybyla AE, MacDonald RJ, Rutter W (1979) Isolation of biologically active ribonucleic acid from source enriched in ribonuclease. Biochemistry; 18:5294–5299
5. Siebert P (1991) RT-PCR. Methods & Applications, Book 1, Clontech Laboratories Inc., Palo Alto
6. Chomczynski P, Sacchi N (1987) Single-step method of RNA isolation by acid guanidinium thiocyanate-phenol-chloroform extraction, Anal Biochem; 162:156–159

Chapter 2.2
Synthesis of cDNA

D. Lassner

2.2.1 Theoretical Background

For detection or quantitation of mRNA by polymerase chain reaction, it is necessary to transcribe mRNA into copy-DNA (cDNA) by a reverse transcriptase (RT) reaction. The quality of a cDNA depends on the integrity of the messenger RNA and the fidelity of the transcription.

Two types of reverse transcriptases (RT, revertase) are commonly used in cDNA synthesis: Avian myeloblastosis virus (AMV) RT and Moloney murine leukemia virus (MMLV) RT.

Like many polymerases, revertase catalyzes more than one reaction. The enzyme embodies two functions in vitro – a polymerase activity and an associated ribonuclease H (RNase H) activity [1].

In general, intrinsic RNase H activity degrades most of the remaining mRNA. In other cases, under non-optimal RT conditions, high RNase H activity decreases the yield of the cDNA obtained. This problem can be minimized by use of recombinant reverse transcriptases lacking RNase H activity.

Intact mRNA contained in the cDNA synthesis mixture may interfere within the subsequent PCR process by competing with synthesized cDNA for the PCR reaction components.

A recombinant DNA polymerase derived from the thermophilic eubacterium *Thermus thermophilus* (Tth pol) was found to possess very efficient reverse transcription activity in the presence of $MnCl_2$ [2]. Many problems typically associated with the high degree of secondary structure in RNA are minimized by using this thermostable DNA polymerase for reverse transcription. The cDNA generated can also be amplified with the same enzyme and in the same tube. A special chelating buffer removes all Mn^{2+} so that the additional Mg^{2+} switches the reverse transcription activity of Tth pol to DNA polymerase activity and starts the PCR process.

Performance of RT reaction and amplification in the same tube is called "single-tube RT-PCR" [3].

Tth pol is ideally suitable for coupled RT-PCR (see chapter 2.4). The cDNA template for RT-PCR is synthesized from a RNA by extension of an annealed primer. There are three ways of cDNA priming:

1. Random Priming
2. Oligo-(dT) Priming
3. Specific Priming

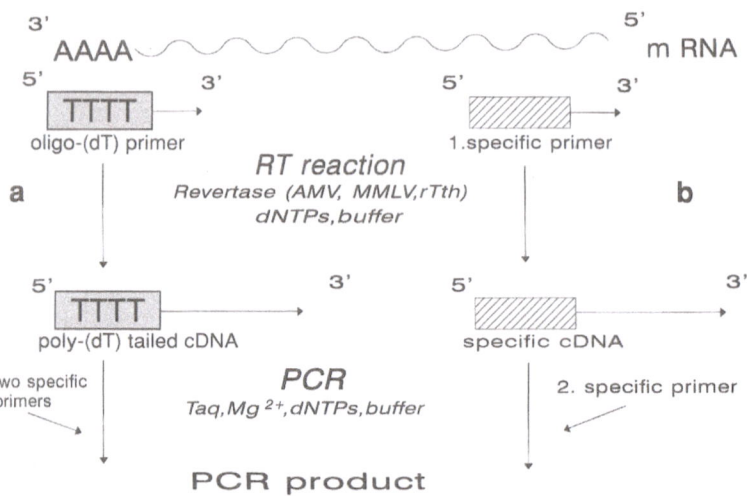

Fig. 2.2-1. Priming of cDNA in RT-PCR. **a** Oligo-(dT) method. Oligo-(dT) primers are annealed to ubiquitary poly A+ tail of mRNA and are extended by reverse transcriptase. **b** Specific method. An specific downstream primer is annealed to the corresponding mRNA strand and extended by revertase

By random priming using random hexamers the entire population of mRNA molecules is converted into a summary of cDNAs of different lengths. This way can be beneficial for the detection of rare mRNAs but is not usual for transcript quantitation by PCR.

The second and third ways are shown schematically in Figure 2.2-1.

The cDNAs of all expressed mRNA are obtained when oligo-(dT) is used as the primer in the RT reaction [4]. The specificity of cDNA synthesis and subsequent amplification is increased by use of the respective specific primer complementary to mRNA.

The mRNA sequence is always identical to the sequence of coding (sense) strand of genomic DNA. The strand serving as the template for synthesis of heterogeneous nuclear RNA (hn RNA) is called the template (antisense) strand and its sequence is complementary to the coding strand of DNA [5].

Upstream primers are identical to the sequence of RNA or the DNA coding strand. Downstream primers are complementary to the desired mRNA and therefore act as starting points for specific cDNA synthesis. Second strand synthesis is important for many applications, for example cloning strategies. Single strand cDNA is sufficient to serve as a PCR template.

2.2.2 Experimental Procedures

For quantitation it is necessary to reverse transcribe mRNA into a cDNA sequence as completely as possible.

Two protocols have been described for the RT reaction using different enzymes. Both are well suited for specific priming but oligo-(dT) priming is not recommended in rTth RT.

Whole RT samples are incorporated after cDNA synthesis in an amplification (coupled RT-PCR, see chapter 2.4).

2.2.2.1 Reverse Transcriptase Reaction with AMV-RT

Purified RNA is reverse transcribed by AMV reverse transcriptase using oligo-(dT) or a specific downstream primer. For quantitation of mdr1-mRNA by parallel amplified external standard cRNA (chapter 3.1) cDNA is specifically primed and the complete RT reaction mixture is introduced into the subsequent PCR assay (see chapter 2.4).

Materials:

1) *5 × RT buffer (250 mmol/l Tris/HCl; pH 8.3, 250 mmol/l KCl, 50 mmol/l MgCl$_2$, 50 mmol DTT, 2.5 mmol/l spermidine) (Promega)*
2) *dNTPs: 10 mM each nucleotide (Promega)*
3) *Down-stream primer (200 ng/µl)*
4) *RNase inhibitor (22 U/µl) (AGS)*
5) *Reverse transcriptase AMV (5 U/µl) (Promega)*
6) *DEPC-H$_2$O*
7) *MicroAmp reaction tubes (Perkin-Elmer)*
8) *GeneAmp 9600 thermal cycler (Perkin-Elmer)*
9) *Mastertips and Biomaster (Eppendorf)*
10) *Eurotips and Varipettes 4810 (Eppendorf)*
11) *Iso-Rack, white (0 °C) (Eppendorf)*
12) *Centrifuge 5415C (Eppendorf)*

Protocol 1: cDNA Synthesis with AMV Reverse Transcriptase

1. All solutions for RT mix are stored at $-20\,°C$. Before use, warm them up to room temperature (except AMV-RT and RNase inhibitor). Vortex and spin all solutions briefly before use.
2. Perform all pipetting steps with a Biomaster pipettor and Mastertips (except for DEPC-H$_2$O).
3. Prepare a 20 µl reaction (to transcribe up to 1 µg RNA) by adding the following reagents:
   - DEPC-H$_2$O   add 16.5 µl
   - 5 × RT buffer   2.0 µl
   - Downstream primer or oligo-(dT) (200 ng/µl)   1.0 µl
   - dNTPs (10 mmol/l each nucleotide)   1.0 µl
   - RNA (up to 1 µg)   X µl

4. Incubate this mixture at 65 °C for 15 min. Place the tube on ice and allow to stay for 5 min. Add the following reagents to the chilled RT mix and incubate for at least 1 h at 42 °C:
   - 5 × RT buffer   2.0 µl
   - AMV reverse transcriptase (5 U)   1.0 µl
   - RNase inhibitor (10 U)   0.5 µl

Prepare a mastermix for all RT reactions by adding necessary volumes of the required solutions to one tube. Vortex and centrifuge at maximum speed to collect sample at the bottom of reaction tube.

The mastermix containing all the insensitive components may be stored at room temperature until aliquoting to each reaction tube and heating to 65 °C. The enzyme mix must be incubated on ice.

Trouble Shooting

Incubation of reaction mixtures at 65 °C removes secondary structures of RNA molecules. Rapid cooling on ice stabilizes the linearized RNA single strand and the enzymes can be added at this time.

Dithiothreitol (supplied within the 5 × RT buffer) is an activator of RNase inhibitor but is thermolabile. Therefore it is necessary to add an aliquot of RT buffer after incubation at 65 °C. In some protocols DTT is added with the enzyme mix.

The amount of added primer is usually 200 ng per RT reaction corresponding to a 20-mer of primer. Scale up or down if primer has a quite different size.

For many protocols 6 units of reverse transcriptase and 10 units of RNase inhibitor are quite sufficient for a standard 20 µl cDNA synthesis reaction. Higher concentrations of the enzymes are not sufficient to increase the cDNA yield.

2.2.2.2 RT Reaction Using rTth DNA Polymerase

cDNA synthesis is limited by stable secondary structures. This problem may be circumvented by increasing reaction temperature and/or using thermostable reverse transcriptase e.g. Tth DNA polymerase. For full activity of Tth pol an optimal reaction temperature of 72 °C and Mn^{2+} ions are required.

Temperature profile of RT reaction using Tth polymerase is comparable with a PCR cycle: starting with linearization of mRNA molecule (at about 70 °C) and cooling to the annealing temperature of the selected primer (about 50 °C) the extension step is performed at 72 °C. Tth RT is very suitable for specific mRNA priming. Oligo-(dT) priming is not recommended because of the relatively low annealing temperatures of this kind of primer. Concentration of primers, buffer and nucleotides are very similar as compared to conventional RT reactions.

Tth DNA polymerase is efficiently used in a coupled reverse transcription reaction where both RT and PCR are performed by the same enzyme [2] (see chapter 2.4). All used reagents are e.g. provided with the "GeneAmp Thermo-stable rTth Reverse Transcriptase RNA PCR Kit" (Perkin Elmer). Tth DNA polymerase can be ordered separately.

The following protocol is used for reverse transcription of mdr-1 mRNA from 50 pg to 1500 ng of total cellular RNA isolated from multidrug resistent cell line CCRF ADR 5000.

Materials:

1) *10 ×rTth Reverse Transcriptase buffer (100 mmol/l Tris/HCl; pH 8.3, 900 mmol/l KCl)(Perkin-Elmer)*
2) *MnCl₂ (10 mmol/l) (Perkin-Elmer)*
3) *dNTPs (10 mmol/l of each nucleotide) (Promega)*
4) *Downstream Primer NP2: 200 ng/µl (sequence: see chapter 2.3)*
5) *rTth DNA polymerase: 2.5 U/µl (Perkin-Elmer)*
6) *DEPC-H₂O*
7) *MicroAmp reaction tubes (Perkin-Elmer)*
8) *GeneAmp 9600 thermal cycler (Perkin-Elmer)*
9) *Mastertips and Biomaster (Eppendorf)*
10) *Eurotips and Varipettes 4810 (Eppendorf)*
11) *Iso-Rack, white (0 °C) (Eppendorf) or ice*
12) *Centrifuge 5415 C (Eppendorf)*

Protocol 2: RT Reaction with rTth Polymerase

1. All solutions for the RT mix are stored at – 20 °C (Store MgCl₂ at + 4 °C, do not freeze). Before use, warm up the vials to room temperature. rTth DNA polymerase should be kept on ice.
2. Vortex and spin all solutions briefly before use.
3. Perform all pipetting steps with Biomaster and Mastertips (except DEPC-H₂O).
4. Pipette following solutions (sufficient for one assay):
   - DEPC-H₂O   add 20.0 µl
   - 10 × rTth Reverse Transcriptase Buffer   2.0 µl
   - MnCl₂   2.0 µl
   - dNTPs   0.4 µl
   - Downstream primer NP2   1.0 µl
   - rTth DNA polymerase   1.0 µl
   - RNA   X µl

Always prepare a mastermix for all samples to get comparable reaction conditions. Keep mastermix and aliquots on ice. Transfer sample tube to preheated thermal cycler (> 60 °C).

5. The temperature profile of the RT reaction with rTth reverse transcriptase is as follows:

    - 72 °C – 5 min
    - 50 °C – 10 min
    - 72 °C – 3 min
    - and finally cooling down to 4 °C

Trouble shooting

The RT reaction performed by rTth polymerase should be considered as one cycle of amplification. Always keep the reaction mix on ice before transferring to the heated cycler.

References

1. Gubler U, Hoffman BJ (1983) A simple and very efficient method for generating cDNA libraries. Gene; 25:263–269
2. Myers TW, Gelfand DH (1991) Reverse transcription and DNA amplification by a *Thermus thermophilus* DNA polymerase. Biochemistry; 30:7661–7666
3. Köhler T, Laßner D, Rost A-K, Leiblein S, Remke H Polymerase chain reaction related approaches to quantitate absolute levels of mRNA coding for the multidrug resistance-associated protein and P-glycoprotein. In: Proceedings of the 2nd International Symposium "Drug resistance in Leukemia and Lymphoma", March 6–8, 1995 Amsterdam, "Advances in Blood Disorders" series, Harwood Academic Publishers (in press)
4. Krug MS, Berger SL (1987) First-strand cDNA synthesis primed with oligo (dT). In: Berger SL, Kimmel AR (eds.). Methods in Enzymology, Vol. 152, Academic Press Inc., New York, pp 316–325
5. Farell RF (ed.) (1993) RNA Methodologies: A Laboratory Guide for Isolation and Characterization. Academic Press Inc., New York

Chapter 2.3

Qualitative RT-PCR: Amplification of Synthesized mdr-1 cDNA

TH. KÖHLER

2.3.1 Theoretical Background

The polymerase chain reaction (PCR) is an in vitro technique allowing the amplification of specific DNA subsequences by simultaneous primer extension carried out by a heat-stable DNA polymerase added to the reaction mixture. Now performed in almost every modern laboratory, PCR has become a powerful and sensitive tool in biomedical research and one of the most widely used techniques in mRNA analysis.

Amplification of mRNA molecules to study gene expression can be achieved by a method that combines both synthesis of cDNA from the RNA template by reverse transcriptase reaction (RT, see chapter 2.2) and PCR. The extremely high sensitivity of these two sequential enzymatic steps gives us the ability to detect extremely rare mRNAs, mRNAs in small numbers of cells or in small amounts of tissue, as well as mRNAs expressed in mixed-cell populations.

An optimized amplification protocol is fundamental for detection of nucleic acids by PCR. Because the basic procedure is very simple the optimization of the PCR reaction is not problematic in the majority of cases and should therefore not be the subject of this chapter (we recommend the reading of literature dedicated mainly to the basic procedure, e.g. 1–3). The requirements of the reaction, deoxynucleotides, DNA polymerase, primers, template, and PCR buffer containing magnesium may generally be used in the recommended concentrations, but the variable parameters, e.g. the cycle program must be adapted to each problem.

Here we will describe a basic protocol used for the amplification of a sub-genomic mdr-1 sequence. The mdr-1 gene product coding for a glycoprotein (P-glycoprotein, pgp-170) that functions as a transmembrane drug-efflux pump is causatively responsible for the complex phenotype named multidrug-resistance (MDR) [4, 5]. This phenomenon is one of the major reasons which limit the efficacy of chemotherapy for several types of cancer in causing the acquired or intrinsic resistance of malignant cells to a wide variety of anti-cancer drugs, such as anthracyclines, epidophyllotoxins, vinca alkaloids etc. Mdr-1 gene transcripts linked to a number of adverse prognostic features, including CD34+ surface phenotype, advanced age etc. are detected with high frequency e.g. in patients suffering from relapsed and high-risk Acute Myelogenic Leukemia (AML). Prospective studies have shown that overexpression of mdr-1 significantly reduces the frequency and duration of complete remissions with conventional induction and postremission therapy [6, 7]. The detection of mdr-1 gene expression by RT-PCR in leukemic cells is therefore a crucial factor in the treatment of leukemia and development of alternative therapeutic strategies to circumvent resistance.

2.3.2 Experimental Procedures

A basic PCR protocol is described such as for amplification of a subgenomic mdr-1 sequence from purified, into cDNA transcribed total RNA isolated from the adriamycin-resistant T-lymphoblastic cell line CCRF ADR5000 (we thank Drs. Volker Gekeler, Byk-Gulden GmbH, Konstanz, and Heyke Diddens, Medical Laser Center Lübeck, for providing us with the cell line) and mononuclear cells from patient samples.

- Simultaneous amplification of a β-actin subsequence as a control for cDNA quality
- PCR product analysis by agarose gel electrophoresis
- Restricted digestion of the amplified DNA fragment
- PAGE electrophoresis and silver staining of restriction fragments

2.3.2.1 Polymerase Chain Reaction (PCR), Basic Protocol

Materials:

1) *cDNA samples (synthesis see chapter 2.2):*
   - *mdr-1 positive control (e.g. cDNA from the human drug-resistant cell line CCRF ADR5000, [8])*
   - *mdr-1 negative control-cDNA*
   - *RT-blank (prepared by addition of H_2O instead of RNA to the RT reaction mixture)*
   - *patient samples: place up to 5 ml whole blood or bone marrow (1:1 diluted with PBS) in a centrifuge tube containing 3 ml of lymphocyte separation medium (density: 1.077 g/ml, Boehringer, Mannheim) and centrifuge for 20 min at 1700 rpm. Wash the pelleted cells with PBS, isolate and reverse transcribe 1 µg RNA aliquots into cDNA.*
2) *10 × PCR-buffer (Perkin-Elmer), see appendix*
3) *dNTP mixture (Promega), see appendix*
4) *β-actin specific amplification primers (both approximately 10 pmol/µl, dissolved in H_2O):*
   - *(+)-primer (Exon 2, nt 1126–1146): 5′ ACG GCT CCG GCA TGT GCA AG 3′*
   - *(−)-primer (Exon 3, nt 1434–1454): 5′ TGA CGA TGC CGT GCT GCA TG 3′*
     predicted amplificate length: 314 bp (genomic DNA), 198 bp (cDNA)
5) *aqua bidest*

Table 2.3-1. Reagents and optimal concentration required for a standard PCR reaction

Reagent	recommended final concentration (50 µl reaction mixture)
Tris buffer	10 mmol/l; pH 8.3–8.4 (25 °C)
gelatin/BSA	0.01 % (w/v)
magnesium chloride	1.5–2.5 mmol/l
potassium chloride	50 mmol/l
non-ionic detergents	0.01 % (v/v) NP40 or Tween 20 (0.1 % Triton X-100)
dNTPs	each nucleotide 0.2 mmol/l
primers	both 20–50 pmol
target DNA	50–100 ng
UDG	0.2–1 unit
DNA polymerase	1–2 units

6) *Uracil-DNA glycosylase (UDG), freshly diluted to 0.1 U/µl with H_2O (Boehringer, Mannheim)*
7) *Taq-polymerase (Perkin-Elmer), freshly diluted to 0.5 U per µl with H_2O*
8) *mdr-1 specific amplification primers (approximately 15 pmol/µl H_2O):*
   - *(+)-primer (NP1)(nt 2419–2440) 5′ AGA TCA ACT CGT AGG AGT GTC 3′*
   - *(–)-primer (NP2)(nt 2690–2708) 5′ GGG CTA GAA ACA ATA GTG 3′ predicted amplificate length (cDNA): 289 bp*
9) *MicroAmp reaction tubes (Perkin-Elmer)*
10) *GeneAmp 9600 thermal cycler (Perkin-Elmer)*

The components of a PCR reaction are readily available from commercial suppliers, e.g. a complete kit is distributed by the Perkin-Elmer Corporation ("Gene-Amp" kit) that is recommended to beginners who are carrying out PCR for the first time. Those who wish to assemble and use their own reagents and solutions will require the chemicals listed in Table 2.3-1.

Protocol 1:   Standard PCR Protocol Adapted to Amplification of a mdr-1 Subsequence

1. Preparation of PCR mastermixes (this step is recommended to reduce the number of pipetting steps and to increase the general reproducibility!) pipette and mix the following solutions:
   - mdr-1 mastermix (sufficient for 8 PCR reactions):
     1) 64 µl dNTPs,2) 40 µl PCR buffer,3) 16 µl primer NP1,4) 16 µl primer NP2,5) 16 µl diluted UDG
   - β-actin mastermix (sufficient for 2 PCR reactions):
     1) 16 µl dNTPs,2) 10 µl PCR buffer,3) 4 µl β-actin (+)-primer,4) 4 µl β-actin (-)-primer,5) 4 µl diluted UDG

Table 3.2-2. Pipetting scheme for routine mdr-1 amplification

tube number	1 µl	2 µl	3 µl	4 µl	5 µl	6 µl	7 µl	8 µl	9 µl	10 µl
H₂O	24	24	24	28	26	26	24	26	26	24
mdr-1 mastermix	19	19	19	19	–	19	19	–	19	19
β-actin mastermix	–	–	–	–	19	–	–	19	–	–
positive cDNA control	4	–	–	–	–	–	–	–	–	–
negative cDNA control	–	4	–	–	–	–	–	–	–	–
RT-blank	–	–	4	–	–	–	–	–	–	–
cDNA patient 1	–	–	–	–	2	2	4	–	–	–
cDNA patient 2 etc.	–	–	–	–	–	–	–	2	2	4
Taq polymerase	3	3	3	3	3	3	3	3	3	3

>> **the danger of contamination from sample to sample begins at this point** <<

2. add 19 µl of the freshly prepared mastermixes and aliquots of each cDNA sample to the desired tubes, respectively, according to Table 2.3-2.

Note: UDG can be routinely used in conjunction with dUTP to eliminate PCR "carry-over" contamination from previous DNA synthesis reactions. If all PCR products synthesized and used in your laboratory contain dUTP instead of dTTP and UDG is added prior to each PCR reaction, the enzyme ensures deglycosylation of any contaminating DNA originating from previous PCRs, thus inhibiting its amplification [9, 10]. We recommend the use of only 0.2 U of the enzyme per sample.

3. After brief centrifugation to deposit all of the fluid at the bottom of the microfuge tubes, place tubes directly into the GeneAmp 9600 thermal cycler.

Note: An overlay of the reaction mixture with mineral oil to prevent evaporation is unnecessary if a thermal cycler with special cover heating (e.g. GeneAmp 9600, Perkin-Elmer) is used.

4. Temperature profile of amplification:
   After a 15 min initial incubation step at 37 °C to allow UDG sterilization perform a 10-min denaturation step at 94 °C to heat-inactivate UDG. After each mixture has reached the following programmed temperature of 72 °C add 3 µl of Taq-polymerase dilution to each vial ("Hot-Start" technique) [11]. Than start the desired cycle program.

Cycle parameters:

- 94 °C, 0:30 min
- 53 °C, 0:30 min

- 72 °C, 0:45 min
- final step: 72 °C, 10:00 min, followed by cooling to 4 °C

Note: An alternative to the "Hot-Start" protocol – a newly developed monoclonal antibody directed against Taq DNA polymerase (TaqStart Antibody, Clontech Laboratories, Inc.) – may be used to control the start of PCR. The antibody was shown to neutralize polymerase activity at room temperature while all reaction components are assembled preventing nonspecific targeting and elongation of primers. After initial "melting" of DNA at 94 °C the antibody is completely denatured and the PCR reaction is allowed to proceed under the proper stringent conditions.

2.3.2.2 Analysis of the PCR Products by Agarose Gel Electrophoresis

The amplified cDNA is identified by the size of the PCR product, which is predicted from the known cDNA nucleotide sequence (available from conventional sequence information data). The PCR product can be further validated by restricted digestion (see section 2.3.2.3) and hybridization with product-specific probes (see Dot blot, chapter 2.5).

Materials:

1) *100 bp ladder (Gibco)*
2) *1 × TAE electrophoresis buffer, see appendix*
3) *4 × sample loading buffer, see appendix*
4) *2 % (w/v) agarose gel (AGS), supplemented with 4 μl/100 ml ethidium bromide from a 10 % (w/v) stock solution*

>>**Ethidium bromide is a powerful mutagen and is moderately toxic. Gloves should be worn when working with solutions that contain this dye!**<<

5) *power supply (e. g. GPS 200/400, Pharmacia or GD 251D, AGS)*
6) *electrophoresis apparatus (e. g. Horizontal DNA/RNA self recirculating Minigel electrophoresis system, AGS)*
7) *Biomaster positive displacement pipettes and tips (Eppendorf)*
8) *Centrifuge 5415C (Eppendorf)*

Protocol 2: Standard Agarose Gel Electrophoresis

1. Remove 5–10 μl aliquots from each PCR reaction tube and transfer in separate 1.5-ml- Eppendorf tubes.
2. Add 1 μl of the nondenaturing bromphenol blue/xylene cyanol loading buffer to each tube, mix and centrifuge briefly.
3. Prepare marker mix: add 2 μl loading buffer to 2 μl of the 100 bp ladder, mix and centrifuge as described.

4. Preparation of an 1.5% agarose gel: boil 1.5 g agarose in 100 ml TAE buffer (see appendix) for about 5 min (e.g. use a microwave), cool to 60 °C and add 4 μl/100 ml ethidium bromide, mix gently but avoid bubble formation, pour the mixture onto the gel carrier, insert the desired combs immediately and allow to cool to room temperature.

Before loading gel slots, introduce 1 × TAE buffer in the electrophoresis apparatus to a level about 1 mm above the gel surface.

>>The electrophoresis buffer is supplemented with 4 μl/100 ml ethidium bromide. Always handle with gloves!<<

5. Place each sample mix carefully in the desired gel slots. Place 2 μl of the marker-mix in the outer slots.
6. Carry out the electrophoresis at 100 V. When the electrophoresis is complete, the stained DNA bands may be examined immediately by transillumination using UV light.

Validation and Documentation

Put the gel on a transilluminator (e.g. Vilber Lourmat TFP-20 M) and photograph with a Polaroid camera DS-34 on Polaroid film 667 at f/8, exposure time

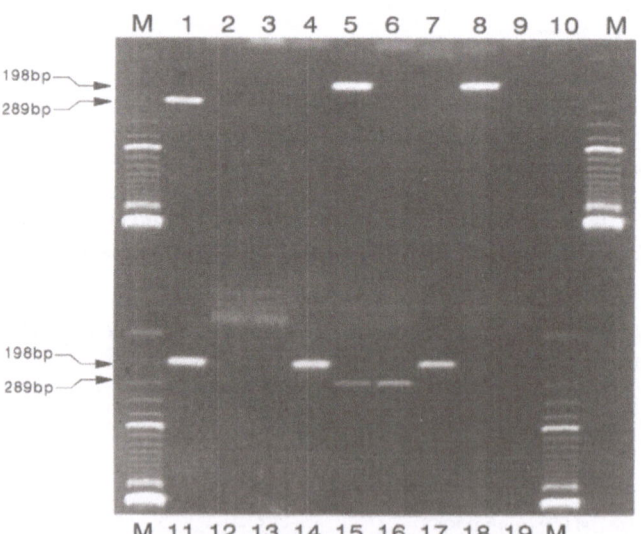

Fig. 2.3-1. Routine analysis of mdr-1 expression in bone marrow samples from patients suffering from leukemia. *Lane M:* 100 bp ladder; *Lane 1:* positive cDNA control (from CCRF ADR5000); *Lane 2:* negative cDNA control (from normal blood donor); *Lane 3:* RT-blank; *Lane 4:* H$_2$O control; *Lanes 5–7:* Pat.W.L. (B-ALL), 2 μl cDNA for β-actin control (198 bp fragment) 2 and 4 μl, respectively for mdr-1 (289 bp fragment), *Lanes 8–10:* Pat. I.V. (T-ALL); *Lanes 11–13:* Pat. A.H. (AML); *Lane 14–16:* Pat. B.S. (AML); *Lanes 17–19:* Pat R.H. (NHL)

1/8 sec. For simultaneous generation of both positive and negative copies, a Polaroid film type 665 is recommended. A typical gel electrophoresis of RT-PCR reaction products derived from leukemia patients is shown in Figure 2.3-1.

>>**Ultraviolet radiation is dangerous, especially to the eyes. To minimize exposure, make sure that the UV source is adequately shielded and the face is protected by a safety mask that efficiently blocks UV light.**<<

Note: A safe way to remove ethidium bromide (a strong carcinogen) from nucleic acid staining buffer waste and for easy disposal is the simple treatment with charcoal or the use of activated carbon filters, e.g. provided by Schleicher & Schüll (Extractor, Ethidium Bromide Waste Reduction System).

Trouble Shooting

The main problem in PCR diagnostics is "carry-over" contamination which may generate confusing and misleading results. Adopting the dUTP/UDG-strategy garantees almost completely contamination-free working. However, unfortunately UDG remains partially active (< 10 %) after an incubation period of up to 30 min at 95 °C. Therefore the manufacturer recommends freezing the PCR product immediately after synthesis. This step is of even more importance for quantitative PCR work (see chapter 3.4).

In addition to such prevention, it is important that control reactions are performed in parallel with the test samples to detect any further contamination. Therefore, as a minimum, the simultaneous amplifications of one positive and three negative checks and an additional check for RNA/cDNA integrity (e.g. amplification of β-actin mRNA, Figure 2.3-1) are recommended.

2.3.2.3. Digesting PCR Products with Restriction Enzymes

Digestion of a given PCR product with appropriate restriction enzymes is a simple and rapid way to check for amplification specificity. The presence of the predicted restriction fragments is therefore proof of this.

Most manufacturers of restriction enzymes have optimized the reaction conditions for their particular preparations and also supply concentrated buffers – therefore one should follow the instructions on the information sheets supplied with each enzyme.

1 – 2 units of restriction enzyme are usually sufficient to cleave up to 1 μg of DNA.

Materials:

1) *Restriction enzyme Rsa I (AGS)*
2) *Restriction enzyme buffer for Rsa I, 10 × concentrated (AGS)*

Table 2.3-3

Enzyme	[μl]	10 × reaction buffer [μl]	H_2O [μl]	predicted length of restriction fragments [bp]
Rsa I	0.5	1.5	3	55, 234
Gsu I	3.5	1.5	–	83, 206

3) *Restriction enzyme Gsu I (MBI Fermentas)*
4) *Restriction enzyme buffer for Gsu I, 10 ×concentrated (MBI Fermentas)*
5) *Thermomixer 5436 (Eppendorf)*

Remove $2 \times 10\,\mu l$ aliquots from the PCR mixture containing the examined fragment and pipette into separate 1.5-ml-tubes labeled with Rsa I (tube 1) or Gsu I (tube 2). Add the following solutions to perform restricted digestion according to Table 2.3-3.

Incubate reaction mix at 37 °C for 1 h. Stop reaction by inactivating the enzyme at 65 °C for 30 min. Store at $0-4$ °C until electrophoresis.

2.3.2.4 Polyacryl Amide Gel Electrophoresis and Silver Staining of Restriction Fragments

Materials:

1) *PAA-gel: CleanGel (Pharmacia) (ready to use): 5 % stacking gel, 10 % resolving gel, 0.5 mm thick, alternatively use a self-prepared gel*
2) *DNA-marker mix: 50, 100, 200, 300, 400, 500, 700, 1000 bp (United States Biochemicals)*
3) *Discontinous buffer system for DNA electrophoresis (Pharmacia) or 1 × TAE buffer (see appendix)*
4) *Multiphor II basic electrophoresis unit (Pharmacia) or other horizontal slab gel unit*
5) *Power supply (1000 V)*
6) *Reagents necessary for silver staining of PAA-gels:*
   * *Fixing-solution: 10 % (v/v) acetic acid*
   * *Silvering-solution: 0.1 % (w/v) $AgNO_3$ in $H_2O + 150\,\mu l$ formaldehyde (for 300 ml solution)*
   * *Developing-solution: 2.5 % (w/v) $Na_2CO_3 + 150\,\mu l$ formaldehyde + a few grains of sodiumthiosulfate*
   * *Stop-solution: 10 % (v/v) acetic acid*
   * *Impregnation-solution: 10 % (v/v) acetic acid + 10 % (v/v) glycerol*

Protocol 3: PAGE Electrophoresis

1. Rehydrate the dry gel and use gel and the electrode strips according to the manufacturer's instructions.
2. Add 5 µl PCR loading buffer to each restriction vial, mix and centrifuge to the bottom. Load 5 µl aliquots of each digested sample carefully into the PAA gel slots. To the right and left lanes apply 1 µl DNA marker mix, and 1 µl uncleaved PCR fragment to a neighboring lane.
3. Run the gel at 600 V for 1 h at 10 °C. At the end of the run remove the gel carefully and transfer to a staining box.
4. Fix 30 min with 250 ml fixing-solution.
5. Wash gel 3 × 2 min with 250 ml H₂O.
6. Stain 20–30 min with 300 ml silvering-solution.
7. Rinse 20 sec with 250 ml H₂O.
8. Develop for 1 min with 100 ml developing-solution under visual control.
9. Stop for 5 min with 250 ml stop-solution.
10. Impregnate for 15 min with 250 ml impregnation-solution followed by air drying at room temperature.

Validation

Measure the distance from the loading well to each of the bands. Plot the $\log_{10}$ of the marker bands against the distance migrated. Use the resulting curve to calculate the size of each restriction fragment and compare with the predicted length obtained from sequence data bases.

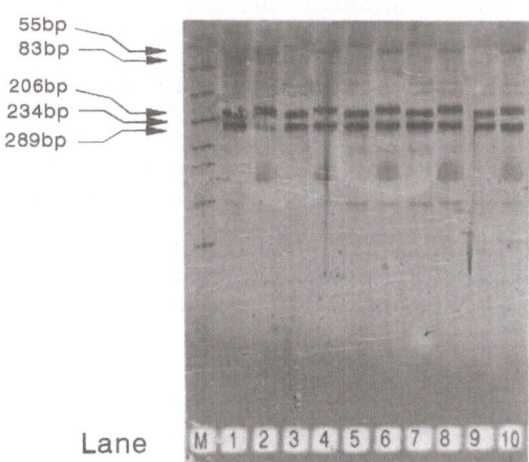

Fig. 2.3-2. PAGE analysis of digested 289 bp mdr-1 fragments, silver-stained. *M* = DNA-marker (USB), *Lanes 1, 3, 5, 7, 9*: digestion with RsaI, *Lanes 2, 4, 6, 8, 10*: digestion with GsuI

Trouble Shooting

If the cleavage is incomplete, digestion for longer periods of time or with excess enzyme does not cause problems unless there is contamination with DNase or exonuclease, but such contaminations are rare in commercially available preparations. If necessary, the digested DNA has to be purified by extraction with phenol:chloroform and once with chloroform and precipitated with ethanol (see section 2.6.2.1 for purification of PCR products).

Use glass or stainless steel dishes for silver staining. Plastic bags may cause elevated background problems.

References

1. Taylor GR (1991) Polymerase chain reaction: basic principles and automation. In: McPherson MJ, Quirke P, Taylor GR (eds.): PCR: A practical approach. Practical approach series, Oxford University Press, New York, pp 1–14
2. Erlich HA (ed.) (1989) PCR technology. Principles and applications for DNA amplification. MacMillan Publishers, London
3. Innis MA, Gelfand DH, Sninsky JJ, White TJ (eds.) (1989) PCR protocols. A guide to methods and applications. Academic Press Inc., Harcourt Brace, Jovanovich Publishers, San Diego
4. Van der Bliek AM, Borst P (1989) Multidrug resistance. In: Advances in Cancer Research, Vol. 52. Academic Press, New York, pp. 165–203
5. McLean S, Hill BT (1992) An overview of membrane, cytosolic and nuclear proteins associated with the expression of resistance to multiple drugs in vitro. Biochim Biophys Acta; 1114:107–127
6. List AF, Spier C, Greer J, Wolf S, Hutter J, Dorr R, Salmon S, Futscher B, Baier M, Dalton W (1993) Phase I/II trial of cyclosporine as a chemotherapy-resistance modifier in acute leukemia. J Clin Oncol; 11:1652–1660
7. Gekeler V, Frese G, Noller A, Handgretinger R, Wilisch A, Schmidt H et al. (1992) Mdr1/ P-glycoprotein, topoisomerase, and glutathione-S-transferase p gene expression in primary and relapsed state adult and childhood leukemias. Br J Cancer; 66:507–517
8. Köhler T, Laßner D, Rost A-K, Leiblein S, Remke H (in press) Polymerase chain reaction related approaches to quantitate absolute levels of mRNA coding for the multidrug resistance-associated protein and P-glycoprotein. In: Proceedings of the 2nd International Symposium "Drug resistance in Leukemia and Lymphoma", March 6–8, 1995, Amsterdam, "Advances in Blood Disorders" series, Harwood Academic Publishers
9. Morales MJ, Gottlieb DI (1993) A polymerase chain reaction-based method for detection and quantification of reporter gene expression in transient transfection assays. Anal Biochem; 210:188–194
10. Bebee RL, Thornton CG, Hartley JL, Rashtchian, A (1992) Contamination-free polymerase chain reaction: endonuclease cleavage and cloning of dU-PCR products. Focus; 14:53–56
11. Ruano G, Brash DE, Kidd KK (1991) PCR: the first few cycles. Amplifications; 7:1–4

Chapter 2.4
Single-Tube RT-PCR

D. Lassner

2.4.1 Theoretical Background

Many PCR protocols for determination of specific mRNAs are based on the synthesis of cDNA in one sample tube and following amplification of an aliquot of RT sample in another reaction tube. This way is preferable when estimating different mRNAs in one RNA/cDNA sample (chapter 2.3). In this case oligo (dT) priming is recommended to obtain cDNA templates of all the desired mRNAs.

In single-tube RT-PCR [1] initially mRNA is reverse transcribed into cDNA (20 µl RT reaction) and subsequently a PCR mix is added to get a final volume of 50 or 100 µl [2]. Reverse transcriptases (AMV, MMLV) are unable to synthesize cDNAs in PCR buffers, whereas Taq DNA polymerase can perform amplification in RT reaction conditions [3]. The concentration of the enclosed amplification mix corresponds to the usual PCR assay including Taq polymerase and primers (see chapter 2.3). The RT reaction solution is considered to be equivalent to amplification solution. The activity of the reverse transcriptase used is destroyed by initial denaturation at 95 °C for 5 min. Emerging amplification products are synthesized by thermostable DNA polymerase.

For incorporation of the whole RT reaction into the PCR process, a specific priming of cDNA is favorable. In this case, only the specific upstream primer must be appended with the amplification solution. The concentration of the added upstream primer should be identical to that of the downstream primer used for cDNA synthesis.

For comparison of two samples, equal volumes of a definite RNA/cDNA sample should be transferred to the RT-PCR reaction vial. Pipetting inaccuracies during apportioning of the RT sample to the PCR reaction mixtures are the main source of sample-to-sample variations. Regarding the exponential accumulation of the yield throughout the amplification process, only slight differences in the starting concentration of the template may lead to quite different quantitation values when PCR is finished. Tube to tube differences can be diminished by incorporation the whole RT sample into the PCR reaction.

2.4.2 Experimental Procedures

These protocols are a direct continuation of cDNA synthesis (see chapter 2.2). The whole RT reaction is used for amplification of the synthesized mdr-1 cDNA.

cDNA synthesis is specifically primed using the corresponding downstream primer NP2 (do not use oligo (dT) primer) (see chapter 2.2).

Perform PCR amplification of the newly synthesized cDNA by adding a 5'-biotinylated second mdr-1 specific primer as prerequisite for quantitation of accumulated product by ELOSA technique (chapter 3.1).

2.4.2.1 Amplification of Single-Stranded cDNA with Taq DNA Polymerase

The cDNA of the RT reaction with AMV-RT is used completely in the following PCR assay. 30 µl of PCR mastermix is added to the RT sample to yield a final volume of 50 µl.

Materials:

1) *$10 \times$ PCR buffer (Perkin-Elmer): see appendix*
2) *dNTPs (1.25 mmol/l dA/G/C/UTP) (Promega)*
3) *Upstream primer NP1 (biotinylated, 200 ng/µl)*
4) *Taq DNA polymerase (diluted to 0.5 U/µl H_2O) (Perkin-Elmer)*
5) *DEPC-H_2O*
6) *MicroAmp reaction tubes (Perkin-Elmer)*
7) *GeneAmp 9600 thermal cycler (Perkin-Elmer)*
8) *Mastertips and Biomaster (Eppendorf)*
9) *Eurotips and Varipettes 4810 (Eppendorf)*
10) *Iso-Rack, white (0 °C) (Eppendorf)*
11) *Centrifuge 5415 C (Eppendorf)*

Protocol 1: Coupled RT-PCR with Taq DNA Polymerase

The amplification reaction is prepared while the RT reaction is still in progress. Thaw frozen components, vortex and spin all solutions briefly before use.

1. Prepare a mastermix for all PCR samples in a new Eppendorf tube: Pipette the following solutions (sufficient for one sample):
   - DEPC-H_2O 18.2 µl
   - $10 \times$ PCR buffer   3.0 µl
   - dNTP   4.8 µl
   - Primer NP1   1.0 µl
   - Taq polymerase   3.0 µl
2. Transfer tubes containing AMV-RT from thermal cycler to a white Iso-Rack or into ice.
   Vortex and spin the mastermix briefly. Pipette 30 µl of mastermix to the tubes containing cDNA (20 µl).

3. Start program and place the PCR samples in the preheated cycler ($> 60\,°C$, "hot start").
   Use the temperature profile listed in chapter 2.3.1.
4. Analyze PCR products by agarose gel electrophoresis (see section 2.3.2.2) and estimate the synthesized DNA by the ELOSA technique (see chapter 3.1).

2.4.2.2 Amplification of mdr1 cDNA with rTth DNA Polymerase

cDNA synthesized with rTth polymerase is used for the following amplification protocol (see chapter 2.2.2.2). A PCR mix consisting of chelating buffer (chelates intrinsic Mn^{2+}) and $MgCl_2$, which is an essential cofactor for DNA polymerase activity, is added. Amplification is performed using essentially the same temperature profile as described above.

Materials: (see also section 2.2.2.2)

1) *$MgCl_2$ (25 mmol/l) (Perkin-Elmer)*
2) *10 × Chelating buffer (100 mmol/l Tris/HCl pH 8.3, 1 mol/l KCl, 7.5 mmol/l EGTA, 50 % (v/v) glycerol, 0.5 % (v/v) Tween 20) (Perkin-Elmer)*
3) *Upstream Primer NP1(200 ng/μl) (sequence see chapter 2.3)*
4) *DEPC-H_2O*
5) *MicroAmp reaction tubes (Perkin-Elmer)*
6) *GeneAmp 9600 thermal cycler (Perkin-Elmer)*
7) *Mastertips and Biomaster (Eppendorf)*
8) *Eurotips and Varipettes 4810 (Eppendorf)*
9) *Iso-Rack, white (0 °C) (Eppendorf)*
10) *Centrifuge 5415 C (Eppendorf)*

Protocol 2:   Single-Tube RT-PCR with rTth DNA Polymerase

Thaw all solutions after pipetting RT reaction with rTth polymerase (see section 2.2.2.2).
   Vortex and spin solutions briefly before use.

1. Prepare mastermix for all PCR samples in an Eppendorf tube.
   Pipette the following solutions (sufficient for one sample):
   - DEPC-H_2O 61 μl
   - $MgCl_2$ 10 μl
   - Chelating buffer   8 μl
   - Upstream primer NP1 (biotinylated)   1 μl
2. Vortex and spin the mastermix briefly. Pipette 80 μl of rTth PCR mastermix to the MicroAmp tubes supplemented with rTth RT sample stored at 0 °C.

3. Start PCR program and place the PCR samples in the preheated thermal cycler (hot start, > 60 °C).
4. Analyze PCR products by agarose gel electrophoresis and ELOSA (see chapter 3.1).

Trouble Shooting

For successful single-tube RT-PCR the concentration of upstream and downstream primers (supplemented with RT reaction) should be identical. The concentration of DNA polymerases could be reduced to about 0.5 U per sample. However, for a robust quantitative PCR system do not limit this component. General precautions should be carried out as described in chapter 2.3.

References

1. Köhler T, Laßner D, Rost A-K, Leiblein S, Remke H (in press) Polymerase chain reaction related approaches to quantitate absolute levels of mRNA coding for the multidrug resistance-associated protein and P-glycoprotein. In: Proceedings of the 2nd International Symposium "Drug resistance in Leukemia and Lymphoma", March 6–8, 1995, Amsterdam, "Advances in Blood Disorders" series, Harwood Academic Publishers
2. Myers TW, Gelfand DH (1991) Reverse transcription and DNA amplification by a Thermus thermophilus DNA polymerase. Biochemistry; 30:7661–7666
3. Aatsinki JT, Lakkakorpi JT, Pietilä EM, Rajaniemi HJ (1994) A coupled one-step reverse transcription PCR procedure of generation of full-length open reading frames. BioTechniques; 16:282–288

Chapter 2.5

Nonradioactive Determination of PCR-Products by Using a DIG Labeled DNA Probe (Dot Blot)

TH. KÖHLER

2.5.1 Theoretical Background

Dot and slot hybridization is a nucleic acid analysis technique in which an excess of labeled probe is hybridized to a target that has been immobilized on a solid support. Currently nucleic acid hybridizations are most frequently performed by radioactive detection methods particularly by using the ^{32}P-labeling of nucleic acids as the preferred technique [1, 2]. Different nonradioactive methods e.g. using probes labeled with biotin [3] or digoxigenin-dUTP (DIG) [4] have been introduced as possible alternatives to radioactive assays enabling even higher sensitivity than ^{32}P-based hybridizations [4].

Nonradioactive hybridization protocols may also combined with previously performed quantitative RT-PCR allowing the detection of synthesized PCR fragments e.g. as described for quantitation of Hepatitis C virus RNA [3]. Whereas dot or slot blot assays were mostly adapted to measurement of various mRNAs by Northern hybridization [1, 2] these techniques may be also used for quantitation of PCR fragments derived from genomic DNA [5] or RT-PCR [3, 6]. Here we describe a simple DIG-based Dot blot protocol which allows both statements about specificity of synthesized PCR products and semiquantitative efforts in determination of the amplification efficiency of a convenient PCR reaction as prerequisite for quantitation of absolute numbers of target molecules of interest.

A synthetic oligonucleotide probe labeled with digoxigenin-dUTP (DIG) at the 5′ end is used for the experiment (for in vitro labeling of DNA-probes with DIG-dUTP see chapter 2.5). After hybridization of the DIG-probe to the target DNA the hybrids can be detected by a subsequent enzyme-linked immunoassay using anti-digoxigenin alkaline phosphatase-antibody conjugates followed by a simple colorimetric detection.

2.5.2 Experimental Procedures

- Nonradioactive detection of PCR-products by Southern hybridization using a DIG-labeled oligonucleotide-probe
- Determination of exponential phase and amplification efficiency of a PCR reaction as a prerequisite for quantitative PCR using internal or external standards

2.5.2.1 Preparation of Dot Blots

Materials:

1) *GB 003 filter paper sheets (102 × 133 mm, Schleicher & Schüll)*
2) *Nytran-Plus nylon membrane, 0.45 μm, positively charged or BA 85 nitro-cellulose membrane (Schleicher & Schüll)*
3) *Minifold Dot-Blot apparatus SRC 96D (Schleicher & Schüll)*
4) *Vacuum unit*
5) *2 × SSC buffer (prepared from the 20 × stock solution, see appendix)*
6) *DNA sample buffer (SB): 20 mmol/l Tris/HCl, 1 mmol/l EDTA; pH 7.6 (20 °C)*
7) *Neutralization buffer(NB): 1.0 mol/l Tris/HCl, 1.5 mol/l NaCl; pH 7.5 (20 °C)*
8) *DNA dilution buffer (DB): 10 mmol/l Tris/HCl, 1 mmol/l EDTA; pH 8.0 (20 °C), 50 μg/ml salmon sperm DNA*
9) *λ-DNA, 20 ng/μl (Boehringer Mannheim)*
10) *Boehringer positive control (B+): linearized and DIG labeled pBR328-DNA, 100 ng/μl, further diluted 1: 100 with DNA-DB*
11) *1 mol/l NaOH*
12) *UV Stratalinker 2400 (Stratagene)*
13) *IsoTherm-System 3880, white Iso-Racks: 0 – 4 °C (Eppendorf) or ice*

Protocol 1: Dot Blot

Sample Preparation

Blot the desired and denatured PCR-samples in duplicate pure, diluted 1:10 and 1:100 with DNA dilution buffer, respectively. To test the reliability of the system spot in addition two known concentrations of DIG-labeled DNA (positive control). Proceed as summarized in Table 2.5-1.

Preparation of Dilutions

Add 1 volume PCR sample or control to 9 volumes of DNA-SB and denature for 10 min at 95 °C. Cool rapidly to 4 °C using white Iso-Racks or ice, bring down condensation by a brief spin. Add an equal volume of 1 mol/l NaOH and incubate 20 min at room temperature, neutralize with 4 volumes of neutralizing buffer and mix gently (Table 2.5-1).
When all samples including the controls are ready for application pipette 495 μl sample aliquots in duplicate to the desired sample wells of the filtration manifold (operate quickly to prevent renaturation!).

Preparation of the Minifold Apparatus

The Minifold filtration apparatus was designed to accept a large number of samples and to deposit the nucleic acids onto blot filters in a fixed pattern to allow the results to be quantitated by scanning densitometry.

Table 2.5-1

[μl]	PCR samples			Controls			
	pure	1:10	1:100	blank	B+	B+ 1:10	λ-DNA
PCR-sample	10	1	10	–	–	–	–
control-DNA	–	–	–	–	1	1	10
DNA-DB	–	9	–	–	4	4	–
DNA-SB	90	90	90	100	45	45	90
incubate 10 min at 94 °C, rapidly transfer to ice							
1 mol/l NaOH	100	100	100	100	50	50	100
leave at room temperature for 20 min, transfer to ice							
neutralization buffer	800	800	800	800	400	400	800

1. Wet a piece of filter paper in $20 \times$ SSC and place it on the filter support plate on the vacuum filtrate chamber ensuring that the cut-off corners leave the registration pins exposed.
2. Prepare nitrocellulose membranes (e.g. BA 85) by presoaking in water and then $20 \times$ SSC. Nitrocellulose membranes must be dried before loading with DNA. Nylon membranes can be used without any pretreatment [7]. Place the membrane on the filter paper sheet matching up to the cut-off corners.
3. Place the sample well plate on the top and clamp the whole Minifold together.
4. Apply the chilled samples quickly.
6. Apply gentle suction to the minifold, continue suction of all the fluid for about 5 min.
7. Remove the filter from the minifold, and allow to dry completely at room temperature.
8. Fix DNA by exposing the side of the membrane carrying the DNA to ultra-violet irradiation (254 nm) of 0.12 J/cm^2 in an UV-linker (e.g. Stratalinker 2400, Stratagene), alternatively a normal transilluminator may be used. For fixation of nucleic acids to nitrocellulose baking at 120 °C for 15–30 min is recommended [7].

>>**Ultraviolet radiation is dangerous, especially to the eyes. To minimize exposure make sure that the UV source is adequately shielded and the face is protected by a safety mask that blocks UV light efficiently.**<<

The filter is now ready to use for Southern hybridization or can be alternatively stored dried and dust-free for several days. After crosslinking cut the filters if desired with a sterile razor blade or scissor.

2.5.2.2 Hybridization of the Blots with a DIG-Labeled Probe

Materials:

1) *Hybridization solution: 5×SSC (from a 20×stock solution!), 0.5% (w/v) blocking reagent (Boehringer DIG-labeling and detection kit), 0.1% (w/v) lauroyl sarcosine (Fluka AG), 0.02% (w/v) SDS (Serva)*
2) *Washing buffer 1: 2×SSC, 0.1% (w/v) SDS*
3) *Washing buffer 2: 0.5×SSC, 0.1% (w/v) SDS*
4) *5'-DIG labeled ssDNA probe:*
   Sequence: 5' GCTGGTTGCA GGCCTCCATT TATAATG 3'
5) *50 ml centrifuge tubes (Greiner)*
6) *Hybridization oven e.g. Hybaid "Mini" (distributed by MWG Biotech)*
7) *Thermomixer 5436 (Eppendorf)*

Protocol 2: Southern Hybridization

1. Adjust the hybridization oven to 53°C 1–2 h before starting the experiment.
2. Label the blot filter to allow proper recoding.
3. Label a 50-ml-centrifuge tube and place the filters carefully in the tube with the side of the membrane carrying the DNA to the inner room.
4. Add 5 ml of the hybridization solution, which has been prewarmed to 53 °C, close the tubes tightly, drill a small hole in the middle of the screw cap (for pressure compensation).
5. Fasten the centrifuge tubes in the intended supports, start rotation, and incubate for 1 h at 53 °C.
6. Before finishing prehybridization, denature the DIG probe for 10 min at 95 °C using a thermomixer (this step is not absolutely necessary when using short single-stranded oligonucleotide probes), cool rapidly on ice to prevent renaturation.
7. After performing prehybridization, remove tubes from the oven and replace the solution with 5 ml prewarmed hybridization solution, add 16 μl denatured probe to give a final concentration of about 40 ng per ml.
8. Hybridize over night at 53 °C under constant rotation.
9. Discard the hybridization solution.
10. Replace the perforated screw cap with a new one, wash the membrane 2×5 min with 10 ml washing buffer 1 at room temperature with constant rotation.
11. Wash the filters meticulously for 2×15 min with 10 ml washing buffer 2 at 53 °C (hybridization oven).
12. Use blots immediately for detection of DNA hybrids or store air-dried for several days.

Note: The adequate hybridization temperature for a used probe reflecting its length and base composition can be calculated according to the following formulae [8]:

- For probes longer than ~ 50 nucleotides:
  $T_m = 81.5 + 16.6 \, (\log M) + 0.41 \, (\% \, G/C) - (500/n) - 0.62 \, (\% \, formamide)$
- For hybrids between oligonucleotides (14–20 bp) and immobilized DNA:
  $T_d = 4 \, (G + C) + 2 \, (A + T)$

where M is the cation concentration in the hybridization solution (mol/l), n is the probe length (bases), A, C, G, T are the nucleotides, T_m the melting temperature, and T_d the temperature at which 50% of the short duplexes dissociate.

2.5.2.3 Detection of DNA-DNA Hybrids

Materials:

1) *Buffer 1: 100 mmol/l Tris/HCl, 150 mmol/l NaCl; pH 7.5 (20 °C)*
2) *Buffer 2: 0.5 % (w/v) blocking reagent (Boehringer) dissolved in buffer 1 (prepare at least 1 h before use and dissolve at hybridization temperature!)*
3) *Buffer 3: 100 mmol/l Tris/HCl, 100 mmol/l NaCl, 50 mmol/l MgCl$_2$; pH 9.5 (20 °C)*
4) *NBT (nitro blue tetrazolium) solution: 75 mg/ml in dimethylformamide*
5) *X-phosphate (5-bromo-4-chloro-3-indolylphosphate, toluidin-salt): 50 mg/ml in dimethylformamide*
6) *TE-buffer, see appendix*
7) *Staining solution (prepare freshly!, all components are contained in the Boehringer kit but are also available separately from different companies) and staining dish*

Protocol 3: Detection of Hybridized Probe with the Alkaline Phosphatase Coupled Anti-Digoxigenin Antibody

Perform all following steps at room temperature under constant shaking, with the exeption of color development.

1. Wash blot membrane briefly (1 min) with 15 ml buffer 1.
2. Block with 10 ml buffer 2 (containing the blocking reagent) for 30 min.
3. Wash briefly with 15 ml buffer 1.
4. Incubate with antibody conjugate: add 2 µl conjugate to 10 ml buffer 1 (i.e. dilution of 1:5000) for 30 min.
5. Discard the antibody solution and wash 2 × 10 min with 15 ml buffer 1.
6. Equilibrate for 2 min in 10 ml of buffer 3.

7. Prepare 30 ml color solution by mixing the following solutions:
   - 135 µl nitroblue tetrazolium salt
   - 105 µl X-phosphate
   - 30 ml buffer 3
8. Remove blot membranes from the tube, transfer to the staining dish and incubate in the dark with the color solution.
9. If clearly visible dots are recognizable (at least 2 hours are necessary, extending the time to as long as 24–48 h is possible) stop the colorimetric reaction with TE-buffer, rinse with destilled water and allow to air dry.

Validation

Measure the color intensity of each dot (a representative dot blot is shown in Figure 3.5-1, Panel A) e.g. by densitometric scanning using a laser scanner GT 6000 (Epson) and validate the signal intensities e.g. by the Gelimage-Software (Pharmacia). The results obtained may be used to determine the efficiency and exponential rate of the amplification process. Plot the $\log_{10}$ of the dot areas against the number of cycles (Figure 3.5-1, Panel B). Calculate the efficiency of the reaction from the slope of the regression straight line and determine the beginning of the plateau phase.

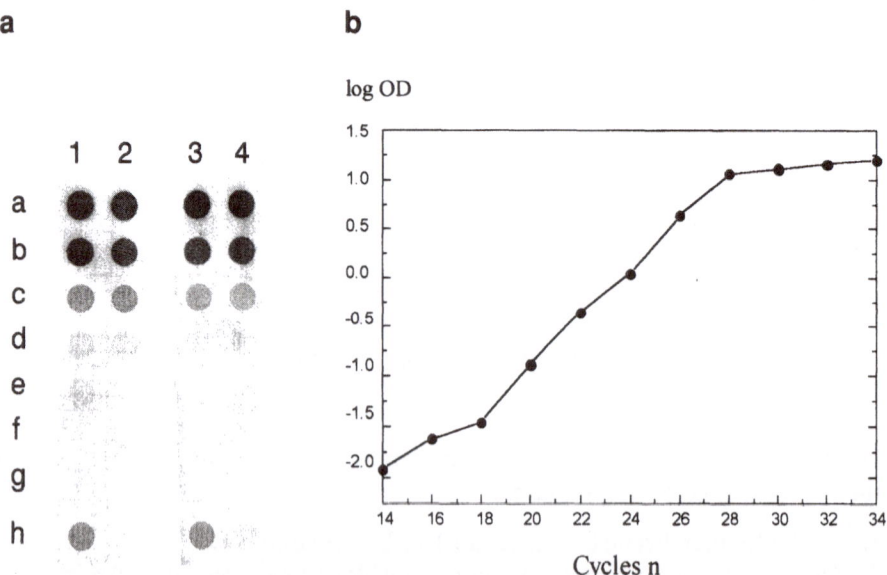

Fig. 2.5-1. Detection of PCR amplified mdr-1 DNA fragments by using a DIG labeled oligonucleotide probe. **a** mdr-1 cDNA amplified by 40 cycles and dotted pure (*a*), 1:10 (*b*), 1:100 (*c*), 1:10³ (*d*), 1:10⁴ (*e*) and 1:10⁵ diluted (*f*); blank (*g*) and Boehringer (+)-control (*h*) 1 pg and 0.1 pg. *Lane 1/2*: 40 ng labeled probe per ml, *lane 3/4*: 20 ng probe per ml. **b** Kinetic analysis of GAPDH amplification reaction using the Dot blot technique and densitometric scanning

References

1. Nooter K, Sonneveld P, Janssen A, Oostrum R, Boersma T, Herweijer H et al. (1990) Expression of the mdr3 gene in prolymphocytic leukemia: Association with cyclosporin-A-induced increase in drug accumulation. Int J Cancer; 45:626–631
2. Gekeler V, Frese G, Noller A, Handgretinger R, Wilisch A, Schmidt H et al. (1992) Mdr1/ P-glycoprotein, topoisomerase, and glutathione-S-transferase p gene expression in primary and relapsed state adult and childhood leukemias. Br J Cancer; 66:507–517
3. Rüster B, Zeuzem S, Roth WK (1995) Quantification of Hepatitis C virus RNA by competitive reverse transcription and polymerase chain reaction using a modified hepatitis C virus RNA transcript. Anal Biochem; 224:597–600
4. Engler-Blum G, Meier M, Frank J, Müller GA (1993) Reduction of background problems in nonradioactive Northern and Southern blot analyses enables higher sensitivity than 32P-based hybridizations. Anal Biochem; 210:235–244
5. Saiki RK, Bugawan TL, Horn GT, Mullis KB, Erlich HA (1986) Analysis of enzymatically amplified β-globin and HLA-DQa DNA with allele-specific oligonucleotide probes. Nature; 324:163–166
6. Murphy Jr GM, Jia X-C, Yu ACH, Lee YL, Tinklenberg JR, Eng LF (1993) Reverse transcription and polymerase chain reaction technique for quantification of mRNA in primary astrocyte cultures. J Neurosci Res; 35:643–651
7. Höltke H-J, Seibl R, Burg J, Schmitz GG, Walter T, Krüger R et al. (1992) The digoxigenin: anti-digoxigenin (DIG) system. In: Kessler C (ed.) Nonradioactive labeling and detection of biomolecules. Springer, Berlin Heidelberg New York, pp. 36–56
8. Wahl GM, Berger, SL, Kimmel AR (1987) Molecular hybridization of immobilized nucleic acids: Theoretical concepts and practical considerations. In: Methods in Enzymology, Vol. 152. Academic Press, New York, pp. 399–407

Chapter 2.6

Nonradioactive Northern Blot Hybridization with DIG-Labeled DNA Probes

A.-K. ROST

2.6.1 Principle and Application of Nonradioactive Northern Blot Hybridization

The originally method of Northern blotting described by Alwine et al. [1] represents a standard tool for determining the size and the amount of a specific transcript within preparations of total or poly (A+) RNA.

First RNA is separated using denaturing agarose gel electrophoresis and is thereafter transferred to a membrane filter. Then the mRNA species of interest is detected by hybridization with a specific DNA or RNA probe.

For quantitative and qualitative evaluation of a target transcript for each sample, a simultaneous hybridization with a probe specific for a housekeeping gene should be performed.

Northern blot hybridization provides a qualitative component to RNA analysis and enables one to check the integrity of the RNA sample. A further advantage of this procedure is that RNA is relatively stable when adsorbed to a solid support [4].

Northern blot hybridization is commonly performed with radioactive-labeled probes. Methods for preparation and application of such probes are well-established but they are limited by cost and hazards associated with radioisotope use and the half-life of the isotope.

Digoxigenin (DIG) based protocols represent a nonradioactive and highly sensitive alternative to labeling with isotopes [2, 3].

DNA or RNA probes are labeled with DIG-11-dUTP/UTP where dUTP/UTP is linked via an 11-atom spacer arm to the steroid-hapten DIG. After hybridization of the DIG-probe to the target RNA on the membrane, the hybrids can be detected by a subsequent enzyme-linked immunoassay using DIG-specific antibodies coupled to alkaline phosphatase. The colorimetric, enzyme-catalyzed detection is performed by using X-phosphate and NBT; Lumigen PPD or CSPD is used as chemiluminescent substrate for alkaline phosphatase.

The method allows the detection of 0.1 pg of homologous RNA within 1–16 h using the colorimetric reaction or detection of 0.03 pg homologous RNA within 15–60 min using the chemiluminescent reaction.

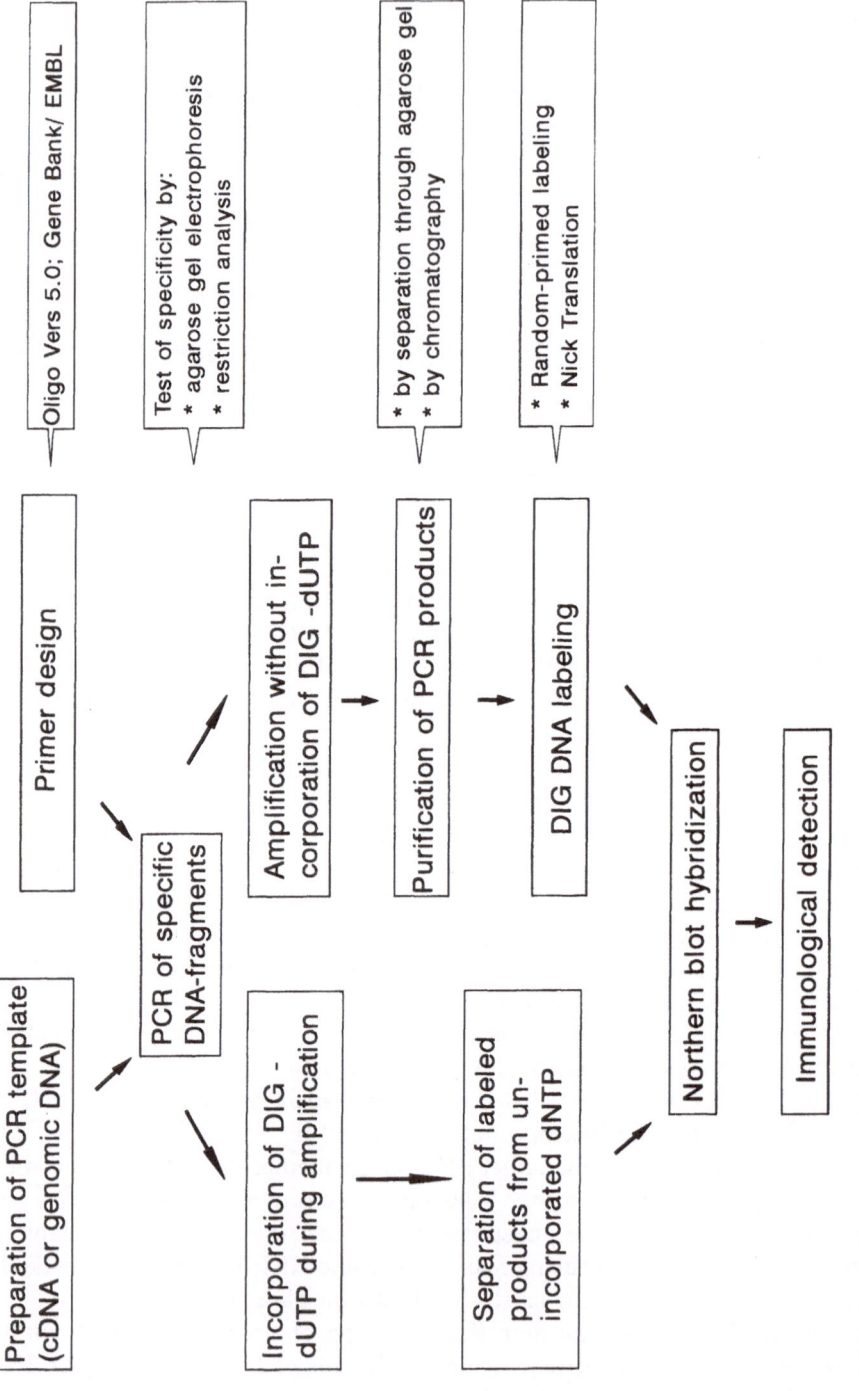

Fig. 2.6-1. Generation of DIG-labeled DNA probes by PCR

2.6.2 Preparation of DIG-Labeled DNA Probes by Using PCR-Generated DNA-Fragments

DIG-labeled RNA probes, synthesized by in vitro transcription of plasmid vectors, and plasmid derived cDNA probes, labeled with DIG-11-dUTP by random priming have become well-established for nonradioactive Northern blot hybridization [5, 6].

Although with such probes highly sensitive Northern blotting for qualitative and semiquantitative evaluation of target mRNA can be performed, the time-consuming preparation of plasmid vectors is a decided disadvantage of this method.

An alternative technique is the generation of DIG-labeled DNA probes by PCR using genomic DNA or reverse transcribed RNA as a template. Two different methods are applicable for synthesizing PCR-generated dsDNA probes (Figure 2.6-1).

First, labeling of PCR-products with DIG can be performed directly during amplification. In this case, the labeling is achieved by application of an 5′ end-labeled primer or by incorporation of DIG-11-dUTP during PCR [7, 8].

In addition, PCR-products can be labeled with DIG after amplification and this is described in the following protocol. First, large numbers of specific DNA target molecules are synthesized by PCR. Thereafter, DIG-11-dUTP is incorporated into purified dsDNA fragments by the subsequently described random-primed labeling or by nick translation.

Hybridization with probes generated by this protocol results in excellent Northern blots with low background and high sensitivity comparable with sensitivity obtained by using ^{32}P-labeled probes if the RNA-DNA hybrids are detected by chemiluminescent reaction [9].

Labeling after amplification is recommended for PCR-products greater than 300 bp in length but for smaller probes the incorporation of DIG-11-dUTP during PCR should be used [8].

Advantages of the generation of DNA probes by PCR include: i) No need to clone the DNA template into a plasmid vector. Therefore the preparation is simple and rapid; specific probes can be synthesized within 1 to 2 days. ii) This method allows high flexibility in probe sequence selection, independent of restriction enzyme site location. iii) The DNA probes are potentially more stable than RNA probes.

2.6.2.1 Synthesis of DNA Fragments by PCR

PCR amplification to generate DNA probes should be performed by an optimized standard protocol (see chapter 2.3) using sequence-specific primers. The amount of DNA synthesized by up to two 50 µl-standard PCR reactions is sufficient to perform the labeling reaction.

2.6.2.2 Purification of DNA Fragments

DNA fragments can be purified from contamination by electrophoretic separation through a TAE agarose gel.

After ethidium bromide staining, DNA bands can be localized by trans-illumination and the DNA of interest can be cut out from the gel. Then the DNA fragments can be extracted by standard protocols [11] or by applying a special preparation kit available from some manufacturers.

We recommend the use of Sephaglas Band Prep Kit (Pharmacia). This procedure gives reproducible yields of about 80–90 % and the procedure may be completed within 30 min. The resulting DNA is ready to use for subsequent enzymatic manipulation.

Before the labeling reaction, purification of PCR products from contamination including primer-dimers, primers and nonspecific products is required to reduce background signals in Northern blot hybridization.

Another way to purify PCR-products contaminated only by primer-dimers and primers is separation by spin-column chromatography, e.g. by using Micro-Spin Colums (Pharmacia) or by Wizard PCR Preps DNA Purification System (Promega).

Materials:

1) *PCR-samples*
2) *Electrophoresis equipment: see chapter 2.3*
3) *Sephaglas BandPrep Kit (Pharmacia)*

Components:

- *Sephaglas BP: Sephaglas BP suspended in a sodium iodide solution buffered with Tris/HCl.*
- *Gel Solubilizer: Sodium iodide solution buffered with Tris/HCl.*
- *Elution buffer: 10 mmol/l Tris/HCl (pH 8.0), 1 mmol/l EDTA.*
- *Wash buffer: Buffered salt solution containing ethanol.*

Protocol 1: Purification of PCR Fragments

The following procedure is designed for extraction of up to 1 µg of DNA from an agarose gel slice weighing up to 250 mg. The volumes and weights in this protocol can be directly scaled up for slices weighing more than this or containing more DNA.

1. Separate the pooled PCR mixture containing the amplified DNA fragment by agarose gel electrophoresis (see chapter 2.3).
2. Localize the DNA bands by transillumination. Excise the DNA band of interest and transfer it into a 1.5 ml microfuge tube. Cut the DNA band as close as possible. Determine the weight of the agarose slice.

Note: A 750 mg plug is the largest that can be extracted in a 1.5 ml microfuge tube.

3. Gel solubilization:
   Add 250 µl of gel solubilizer (250 µl minimum, or 1 µl for each mg of agarose), vortex vigorously, and incubate the tube at 60 °C for 5–10 min until the agarose slice is dissolved. Cool the solution to room temperature.
4. DNA binding:
   Vortex the flask containing the Sephaglas BP to yield a uniform suspension, add 5 µl of the suspension for each estimated µg of DNA to the dissolved gel, vortex and incubate 5 min at room temperature with continuous shaking. Centrifuge at high speed for 1 min in a microfuge, and remove the supernatant without disturbing the pellet.
5. DNA purification:
   Add 40 µl of wash buffer (or 8 × the volume of added Sephaglas BP). Resuspend the pellet completely by vortexing and spin at high speed for 1 min in a microfuge. Remove the supernatant, taking care not to disturb the pellet. Repeat this step twice.
6. Drying:
   Invert the tube and place it on a paper towel on the bench top. Allow the pellet to air-dry for 10 min.
   Never dry under vacuum!
7. DNA recovery:
   Add 10 µl of elution buffer or sterile water (10 µl minimum or 0.5 volumes of added Sephaglas), vortex gently to resuspend the pellet and incubate for 5 min at room temperature with periodic agitation. Centrifuge at high speed for 1 min in a microfuge. Carefully remove the supernatant and place it into a new microfuge tube. Repeat this elution step once more to obtain a better yield of DNA.
8. Estimate the concentration of purified DNA by spectrophotometry (see chapter 2.1) and examine the quality by agarose gel electrophoresis of an aliquot of preparation (see chapter 2.3).

Trouble Shooting:

Recovering the DNA at 60 °C is recommended to increase the DNA yield.
   The purified DNA can used directly for the labeling reaction without further purification, but be sure that the matrix is thoroughly washed and air-dried before DNA elution. Contaminations with iodide or ethanol in the final elution step may lower the DNA recovery and inhibit enzymes used for subsequent reactions.

2.6.2.3 DIG-Labeling of DNA-Fragments by Random Priming

Random priming [12] is used to generate probes from denatured, closed circular DNA or denatured, linear dsDNA but completely linearized DNA is more efficiently labeled than circular DNA.

This method allows labeling of 10 ng up to 3 µg of DNA per standard reaction. The yield of labeled DNA is a function of the amount and purity of the DNA template and the duration of incubation at 37 °C (Table 2.6.-1).

The labeling reaction occurs within 1 h and results in incorporation of DIG-11-dUTP every 20–25 nucleotides into the newly synthesized DNA.

Table 2.6-1. Amount of synthesized DNA in dependence on amount of template DNA and on duration of incubation

Amount of template DNA per labeling reaction	10 ng	30 ng	100 ng	300 ng	1000 ng	3000 ng
Amount of synthesized DIG-labeled DNA						
– after 1 hour-	15 ng	30 ng	60 ng	120 ng	260 ng	530 ng
– after 20 hours-	50 ng	120 ng	260 ng	500 ng	780 ng	890 ng

The size of the DIG-labeled DNA fragments obtained by random primed DNA labeling depends on the amount and size of the template DNA, e.g. using linear pBR328 DNA as template the size of labeled probes is in the range of 200–2000 bp with a peak around 700 bp.

The labeled DNA can be stored frozen at –20 °C indefinitely.
(Source: Boehringer data sheet for DIG-labeling and detection Kit)

Materials:

1) *DNA template: purified PCR products; dissolved in a maximum volume of 15 µl DEPC-treated H$_2$O.*
   It is possible to vary the amount of DNA template from 10 ng – 3 µg (see above). We recommend the use of about 500 – 800 ng per reaction.
2) *Hexanucleotide mixture, 10 conc.; 62.5 A$_{260}$ U/ml random hexanucleotides, dissolved in 500 mmol/l Tris-HCl, 100 mmol/l MgCl$_2$, 1 mmol/l DTT, 2 mg/ml BSA; pH 7.2 (20 °C); (Boehringer, Mannheim)*
3) *DIG DNA labeling mixture, 10 × conc.; 1 mmol/l dATP, dCTP and dGTP, 0.65 mmol/l dTTP, 0.35 mmol/l DIG-11-dUTP; pH 7.5 (20 °C); (Boehringer, Mannheim)*
4) *Klenow enzyme labeling grade, 2 U/µl*
5) *DEPC-treated H$_2$O*
6) *EDTA, 0.2 mol/l; pH 8.0 (20 °C)*
7) *LiCl, 4 mol/l*
8) *Ethanol, 100 % and 75 % (v/v); (– 20 °C)*
9) *TE-buffer (see appendix)*

Protocol 2: DIG-labeling of dsDNA

Pipette the DNA template into a sterile microfuge tube and denature for 10 min at 95 °C. Thereafter chill quickly on ice. Complete denaturation is essential for efficient labeling when using ds DNA.

Add the following reagents to the chilled microfuge tube:

2 µl hexanucleotide mixture and 2 µl DIG DNA labeling mixture; bring the volume up to 19 µl with DEPC-treated H_2O, add 1 µl Klenow enzyme.

Mix the sample well by vortexing, centrifuged briefly and incubated over night at 37 °C.

At the end of incubation, add 2 µl EDTA to stop the labeling reaction.

1. Precipitate the labeled DNA with 2.5 µl LiCl (0.1 volume) and 75 µl prechilled (– 20 °C) 100 % ethanol (2.5 volume). Mix well by vortexing and leave the tube for at least 30 min at – 70 °C or 2 h at – 20 °C.
2. Centrifuge at 14 000 g for 15 min at 4 °C and discard the supernatant.
3. Wash the pellet with about 500 µl cold 75 % ethanol, centrifuge again for 5 min.
4. Decant the 75 % ethanol and dry on air or under vacuum (avoid complete dryness because this impedes resuspension).
5. Dissolve at 37 °C in 50 µl TE buffer for about 30 min.
6. The probe can be stored at – 20 °C for up to one year.

2.6.2.4 Estimating the Yield of DIG-Labeled DNA

Before a newly synthesized DNA probe is used for Northern blot hybridization it is recommendable to estimate the yield of DIG-labeled DNA and confirm the success of labeling reaction by comparison with a labeled control DNA.

Materials:

1) *Labeled control DNA; linearized pBR328 DNA, labeled with DIG according to standard protocol, containing 1 µg template DNA and approximate 260 ng synthesized labeled DNA per 50 µl. (Boehringer, Mannheim)*
2) *DIG-labeled DNA probe*
3) *DNA-dilution buffer (50 µg/ml herring sperm DNA, 10 mmol/l Tris/HCl, 1 mmol/l EDTA; pH 8.0 (20 °C))*
4) *Nylon membrane, positively charged (Boehringer, Mannheim)*

All reagents and solutions necessary for colorimetric or chemiluminescent detection are summarized in section 2.6.5.

Protocol 3: Yield of DIG-DNA Labeling

Perform all dilution steps by using DNA-dilution buffer.

1. Dilute the control DNA 1:5 to a final concentration of 1 ng/μl.
2. Use an aliquot of labeled probe and dilute to an equal concentration. The ratio of dilution should be estimated by using Table 2.6.-1.
3. Prepare 10-fold serial dilutions from both DNAs to obtain the following final concentrations: 100 pg/μl; 10 pg/μl; 1 pg/μl; 0.1 pg/μl; 0.01 pg/μl.
4. Spot 1 μl of each dilution onto a membrane.
5. Fix the DNA onto the membrane by UV-crosslinking and baking (see section 2.6.3.2).
6. Detect the "Spots" corresponding to colorimetric or chemiluminescent detection protocol (see section 2.6.5)
7. Compare the spot intensities of probe and control dilutions and estimate the concentration of labeled probe.

Trouble Shooting:

If the DNA probe is not labeled efficiently, increase the time of labeling reaction (up to 20 h) and be sure to denature the DNA template completely prior to labeling.

Depending on the used purification method, it may be necessary to repurify the DNA by phenol/chloroform extraction and ethanol precipitation (see chapter 1.3).

2.6.3 Preparation of Northern Blots

2.6.3.1 RNA Electrophoresis Through Denaturing Agarose Gels Containing Formaldehyde

Before RNA is transferred to the solid support for Northern blot hybridization the RNA samples have to be size-fractionated by agarose gel electrophoresis.

This method is also applied to determine the integrity of RNA samples which should be examined by other techniques e. g. by PCR.

Electrophoretic separation of RNA requires denaturating conditions to guarantee the migration of RNA molecules only with respect to their molecular weights, because nondenatured RNA frequently develop secondary structures. Such "hairpins" influence the electrophoretic behavior of RNA molecules decisively.

Several denaturants are applicable for agarose gels, including formaldehyde [13, 14] glyoxal/DMSO [15], and methylmercuric hydroxide [16].

Because of its extreme toxicity, methyl-mercuric hydroxide is not recommended for routine use [4].

We prefer the application of formaldehyde as a denaturing agent. The shorter running time of formaldehyde gels compared to glyoxal gels is advantageous. Moreover, recirculation of the running buffer, necessary for glyoxal/DMSO-technique, is not required.

However, Northern blots prepared from formaldehyde denatured RNA samples appear a little less sharp in comparison to RNA samples which have been fractionated through gels containing glyoxal/DMSO [11].

>>To prevent contamination with RNAse, always use gloves, autoclaved pipettes, and vessels. All buffers and solutions must be treated with DEPC (see appendix)!<<

Materials:

1) *RNA-samples:*
    Dissolve a defined quantity of RNA (total cellular or poly (A+) RNA) in DEPC- treated H$_2$O. (The volume should not exceed 4.5 µl per sample when using the following standard protocol.)
    Up to 20 µg total RNA can be separated per lane. For examination of low abundance mRNAs the application of 2–5 µg poly (A+) RNA is recommended.
2) *1% agarose gel containing formaldehyde*
    Preparation of the gel: (The volumes given are suitable for preparation of a 7 × 8 cm minigel, 0.5 cm thick.)
    Boil 0.5 g agarose in 36.7 ml DEPC-treated H$_2$O. Cool the gel to 60 °C, and add 5 ml 10 MOPS and 8.3 ml 37% formaldehyde (1/6 ratio). Mix well and pour it onto the gel bed. Allow the gel to set for at least 30 min.
3) *Running buffer: 1 × MOPS; prepared from a 10 × MOPS stock solution using DEPC- treated H$_2$O (see appendix)*
4) *100% formamide*
5) *37% (v/v) formaldehyde; pH > 4 is essential*

>>Formaldehyde vapor is toxic. Solutions containing formaldehyde should be prepared under a chemical hood and electrophoresis tanks containing formaldehyde solutions should be kept covered whenever possible.<<

6) *10 × MOPS buffer (see appendix)*
7) *Formaldehyde gel-loading buffer (50% (v/v) glycerol; 1 mmol/l EDTA, pH 8.0 (20 °C); 0.5% (w/v) bromophenol blue; DEPC-treated)*
8) *Ethidium bromide (1 mg/ml)*

>>Ethidium bromide is a powerful mutagen. Gloves should be used when working with solutions containing this dye.<<

9) *DEPC-treated H$_2$O*
10) *Electrophoresis chamber (Treat the apparatus for at least one hour with 0.1 N NaOH and rinse with DEPC-H$_2$O before using)*
11) *Power supply*

Protocol 4: Denaturing RNA Electrophoresis

Sample Preparation

1. Prepare each sample for electrophoresis by using the following pipetting scheme (reagent in µl per sample):
   - dissolved RNA*   4.5
   - 100% formamide   10.0
   - 37% formaldehyde   3.5
   - 10 × MOPS   2.0
   - ethidium bromide   0.5
2. Mix the samples well by vortexing and centrifuge for 5 to 10 s. Incubate at 65 °C for 15 min and chill thereafter on ice. After cooling centrifuge for 5 to 10 s to deposit all fluid at the bottom of tube and place on ice immediately.
3. Add 2 µl of gel-loading buffer to each sample, after vortexing centrifuge briefly and chill on ice.

Sample Application and Electrophoresis:

1. Prerun the gel at 5 V/cm for 5 min. Load the samples onto the gel. The gel should not be submerged in running buffer during loading to avoid spillage of RNA.
2. Run the gel at 15 V/cm for 5 min. If samples have entered the gel for 2–3 mm submerge the gel with running buffer and continue electrophoresis at 5 V/cm (2–3 h; bromphenol blue has then migrated halfway down the gel).

Analysis

1. After electrophoresis ethidium bromide stained RNA samples may be examined by ultraviolet illumination and documented by photography (see chapter 2.3).
2. The separated RNA can now be estimated for its integrity:
   By application of eucaryotic total RNA the 18S and the 28S rRNA should appear as two sharp bands whereby the intensity of 28S rRNA band is twice as high as that from 18S rRNA; the 5S and the 5.8S rRNA bands are manifested on leading edge of gel together with tRNA. The significantly lighter smear between, above and below rRNA represents the mRNA components of an intact sample. (Figure 2.6-2)

Trouble Shooting:

If rRNA does not appear as sharp bands either the gel solutions were not mixed adequately or the pH of the formaldehyde was too low. Another explanation

---

* If the sample volume is < 4.5 µl use DEPC-treated H_2O to bring to final volume.

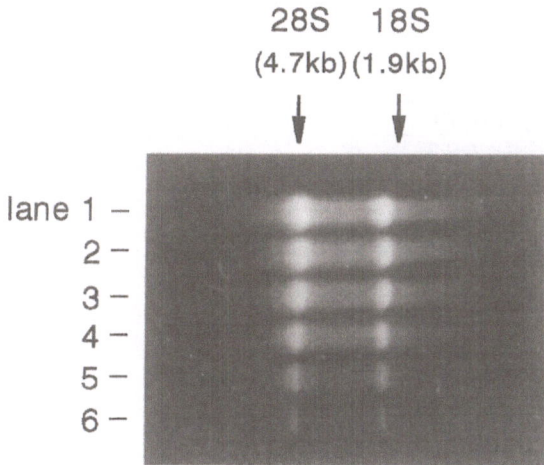

Fig. 2.6-2. Electrophoresis of total RNA samples isolated from human skin fibroblasts through 1% agarose gel containing formaldehyde.
lane: 1 – 15 μg; 2 – 10 μg; 3 – 7.5 μg; 4 – 5 μg; 5 – 2.5 μg; 6 – 1 μg RNA loaded

could be that the sample volume loaded into each slot was too high (for a slot 5 × 1 mm we recommend not to exceed a maximum sample volume of 30 μl).

As an alternative to the protocol described, RNA samples can also be stained after electrophoresis. Cut the lanes to be stained from the gel, and incubate in a solution containing 0.5 mg ethidium bromide/ml in 0.1 mol/l ammonium acetate for 30 – 45 min [11].

2.6.3.2 Capillary Transfer of Denatured RNA to a Nylon Membrane

In addition to vacuum transfer and electroblotting, capillary transfer is a widespread method for absorbing electrophoretically separated RNA from agarose gels to membrane filters. By this technique, buffer is drawn from a reservoir up through the gel into a stack of dry absorbent material. In this way, the RNA is eluted from the gel by the resulting stream of buffer and is deposited on a piece of membrane filter, which is placed on top of the gel. A weight applied to the top of the absorbent material guarantees the required tight connection between the several layers of transfer system (Figure 2.6-3).

Materials:

1) *Nylon membrane, positively charged (Boehringer, Mannheim)*
   Cut a piece of membrane so that it is 2 – 3 mm smaller in both dimensions than the gel.

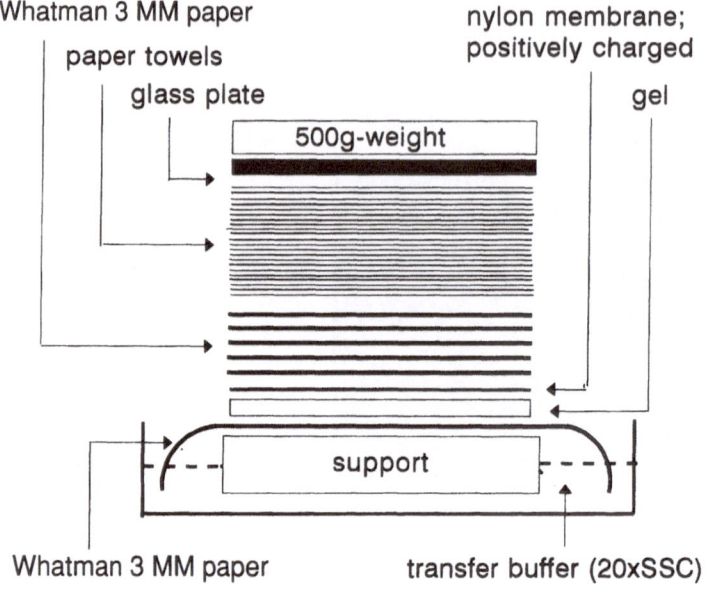

Whatman 3 MM paper

paper towels

glass plate

nylon membrane; positively charged

gel

500g-weight

support

Whatman 3 MM paper   transfer buffer (20xSSC)

Fig. 2.6-3. Principle of capillary transfer of nucleic acids from agarose gels to solid supports

Note: Nylon membranes are preferred because they show significantly higher binding capacity and mechanical resistance than nitrocellulose. In addition, these membranes enable chemiluminescent detection of DIG-labeled nucleic acid molecules and a repeated hybridization without loss of signal.

2) *3 MM paper (Whatman) and paper towels*
3) *Transfer buffer: 20 × SSC (see appendix)*
4) *DEPC-treated H_2O*
5) *Dish (as buffer tank) and gel support (Treat both for at least 1 h with 0.1 N NaOH and rinse with DEPC-H_2O before using.)*
6) *glass plate; 500 g weight*
7) *UV-Stratalinker 2400 (Stratagene) or 302-nm light emitting transilluminator*
8) *Thermostat, 120 °C*

Protocol 5: Capillary Transfer

1. Remove formaldehyde from gel by soaking in DEPC-treated H_2O for 20 min. Thereafter discard the H_2O and soak the gel in 20 × SSC for 45 min. (Both solutions should be changed two to four times during incubation.)
2. Place the gel support within the dish. Fill with 20 × SSC. The level should reach the top of the support.
3. Prepare a piece of 3 MM paper that is longer and wider than the gel. (This piece of paper will act as contact between gel and buffer reservoir.) Wet the

piece of paper with 20 × SSC and place it on the support in such a way that both ends hang into the buffer reservoir, avoid getting air bubbles between paper and support.

4. Place the gel on the 3 MM paper and smooth out all air bubbles.
5. Add 1 to 2 ml 20 × SSC to the surface of the gel and place the dry membrane on the gel. Mark the membrane asymmetrically for proper recoding after transfer. (Make sure that all air bubbles were removed between membrane and gel. Do not move the membrane following application on gel.)
6. Surround the gel with Parafilm to prevent direct flow of liquid from reservoir to absorbent material.
7. Cut 3–8 pieces of 3 MM paper and a stack of paper towels (5–8 cm high). Both should be 2–3 mm smaller in both dimensions than the membrane to avoid an incomplete transfer.
8. Wet a piece of 3 MM paper with 20 × SSC and put it on the membrane. Place the additional pieces of 3 MM paper and the paper towels on the top of the membrane.
9. Put a glass plate and a 500 g weight on the top of the stack. Allow transfer of RNA to proceed at 4 °C overnight (about 16–18 hours).
10. If the transfer is finished, carefully peel the membrane from the gel. Soak the membrane briefly in 5 × SSC to remove any agarose pieces, and put the membrane onto a sheet of 3 MM paper.
11. Immobilize the RNA on the damp membrane by UV-crosslinking and baking at 120 °C for 30 min.
    - By application of a calibrated UV-light source (e.g. Stratalinker, Stratagene) an UV-exposure of 120 mJ/cm² (function: "Autocrosslinking") is recommended. Immobilization with a 302-nm light emitting transilluminator is performed by an exposure time of 3 min.
    - For immobilization of RNA to a filter membrane the instructions of manufacturer should be followed.

The membrane can be used directly for hybridization or stored for later use.

Trouble Shooting:

It is also possible to perform capillary transfer at room temperature for 4 h, but an overnight transfer is more efficient.

To avoid incomplete transfer by capillary elution, the subsequent instructions should be followed:

- The gel should consist of a maximum of 1.2 % agarose and should not exceed 0.5 mm thickness.
- The top weight should be in accordance with gel size to avoid a gel collapse (500 g for a minigel and a maximum of 1250 g for 20 × 20 cm gels).
- The presence of ethidium bromide may result in a reduced transfer efficiency [4, 12]. Therefore the RNA samples should be prepared in duplicate for visualization by ethidium bromide staining.

2.6.4 Northern Blot Hybridization with DIG-Labeled DNA Probes

One of the major problems in applying nonradioactive labeling and detection in Northern blot analysis is the limited sensitivity caused by a nonspecific background.

Background signals are influenced decisively by hybridization conditions (temperature and time as well as composition of prehybridization and hybridization solution). For DNA-RNA hybrids, the protocol published by Church and Gilbert [17] is often recommended [7], but a significant suppression of unspecific signals in connection with increasing sensitivity is achieved by a modification of this protocol according to Engler-Blum et al. [5]. This method is characterized by a general hybridization temperature of 68 °C and a modified composition of hybridization solution. BSA is replaced by 0.5 % (w/v) blocking reagent and the SDS concentration is increased to 20 % (w/v).

Materials:

1) *DIG-labeled DNA probe; dissolved in TE-buffer; pH 8.0 (20 °C) (For preparation see section 2.6.2).*
2) *Prehybridization solution: 0.25 mol/l Na_2HPO_4 (use a 1 mol/l stock solution, pH 7.2 (20 °C), see appendix); 1 mmol/l EDTA; 20 % (w/v) SDS; 0.5 % (w/v) blocking reagent; prepare solution with DEPC-treated H_2O and autoclave it. Store the solution at 4 °C or – 20 °C subsequently.*
   Use 20 ml of prehybridization solution per 100 cm² of membrane.
3) *The hybridization solution consists of prehybridization solution and the denatured DNA-probe. We recommend a probe concentration of about 50 ng/ml hybridization solution. Use 2.5 ml of hybridization solution per 100 cm² of membrane.*
4) *Washing buffer: 20 mmol/l Na_2HPO_4 (use a 1 mol/l stock solution, pH 7.2 (20 °C), see appendix) 1 mmol/l EDTA; 1 % (w/v) SDS*
5) *Hybridization oven (e.g. Mini oven, Hybaid Co.)*

Protocol 6: Hybridization

1. Perform all the following steps in sterile 50 ml plastic tubes. The given volumes are calculated for a membrane size of 100 cm². Alter the volumes according to the used membrane size.
2. Warm up the hybridization oven and prehybridization solution to 68 °C. Because of the high SDS content, allow the solution to incubate 2 – 3 h.
3. Add 20 ml of prewarmed prehybridization solution to the membrane and incubate at 68 °C in the hybridization oven for one hour.
4. After prehybridization replace the solution with 2.5 ml prehybridization solution and add the denatured probe. Incubate the membrane at 68 °C overnight (for at least 16 hours).

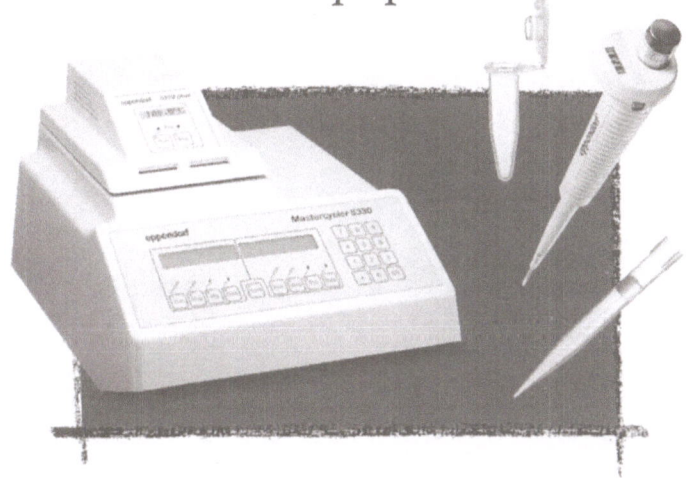

Double stranded DNA probes must be denatured before use!
Incubate the DNA probe at 95 °C for 10 min and chill quickly on ice.

>>Do not allow the membrane to dry between prehybridization and hybridization!<<

5. Discard the hybridization solution and wash the membrane for 3×20 min at 65 °C with 20 ml of washing buffer. It is important to preheat the washing solution to 65 °C!
    This rigorous washing results in the removal of all the unbound probe from the membrane.

The DNA-RNA hybrids can be now detected immediately or alternatively the membranes can, after air drying, be stored for later detection.

Note: The decanted hybridization solution can be reused for further hybridizations. The solution can be stored at – 20 °C for up to one year. For reuse, denature the probe at 68 °C for 10 min and add the solution as previously described.

2.6.5 Immunological Detection of DNA-RNA Hybrids

DIG-labeling enables the visualization of the DNA-RNA hybrids by formation of insoluble color precipitate directly on the membrane, or by chemiluminescence.

Materials:

1) *Buffer A: 100 mmol/l maleic acid, 150 mmol/l NaCl; pH 7.5 (20 °C)*
2) *Buffer B: 1 % (w/v) blocking reagent (Boehringer, Mannheim) in buffer A; prepared from a stock solution containing 10 % (w/v) blocking reagent in buffer A (autoclave the stock solution and store it at 4 °C)*
3) *Buffer C: 100 mmol/l Tris/HCl, 100 mmol/l NaCl, 50 mmol/l MgCl$_2$; pH 9.5 (20 °C)*
4) *DEPC-treated H$_2$O*
5) *Anti-DIG-AP conjugate; polyclonal sheep anti-digoxigenin Fab-fragments, conjugated with alkaline phosphatase, 750 U/ml; store at 4 °C. (Boehringer, Mannheim)*

The following additional materials are required for colorimetric detection:

6) *NBT solution; 75 mg/ml in 70 % (v/v) DMFA; store at – 20 °C.*
7) *X-phosphate solution; 50 mg/ml toluidium salt in 100 % (v/v) DMFA; store at – 20 °C.*
8) *TE-buffer (see appendix)*

The following additional materials are required for chemiluminescent detection:

9) *Buffer A$_{Tween}$: 100 mmol/l maleic acid, 150 mmol/l NaCl; pH 7.5 (20 °C); 0.3 % (v/v) Tween 20*

10) *One of the following chemiluminescent substrates:*
   - *Lumigen PPD; 10 mg/ml disodium salt, (23.5 mmol/l)*
   - *CSPD; 11.6 mg/ml disodium salt, (25 mmol/l)*
   Both substrates should be stored at 4 °C, protected from light.
11) *X-ray film (e.g. Hyperfilm ECL, Amersham)*

Protocol 7: Colorimetric or Chemiluminescent Detection

Perform the following incubations at room temperature in a 50 ml sterile plastic tube or carry out the experiment in a tray or bag. Except for the color reaction, all incubation steps require shaking or mixing. The given volumes are calculated for a membrane size of 100 cm². Alter the volumes according to your membrane size.

Colorimetric Detection with NBT and X-Phosphate

1. Transfer the membrane into a new tube and wash briefly with DEPC-treated H_2O and then with 20 ml of buffer A for 1 min.
2. Incubate with 20 ml of buffer B for 45 min.
3. Dilute anti-DIG-AP conjugate 1:5000 to a final concentration of 150 mU/ml in buffer B.
4. Discard the solution and incubate with 15 ml of freshly prepared antibody-solution for 30 min.
5. Transfer the membrane to a new tube to avoid increasing background signals caused by antibodies bound to the tube wall.
6. Wash 2×15 min with 20 ml of buffer A to remove unbound antibody.
7. Pour off buffer A and equilibrate with 20 ml of buffer C for 2–5 min.
8. Prepare color solution: Add 45 µl NBT solution and 35 µl X-phosphate solution to 10 ml buffer C.

Incubate the membrane with prepared color solution at room temperature in a sealed plastic bag or in a tray protected from light.

The color precipitation starts within a few minutes and the reaction is usually complete within 12–16 h. Do not shake or mix while color is developing!

When the desired bands are detected, stop the reaction by washing for 5 min with 20 ml of TE-buffer and with H_2O for 1 min.

Membranes thereafter may be dried at room temperature. Color fades upon drying but can be revitalized by wetting with buffer D. The membrane can be stored over a long period in a bag in TE-buffer with retention of the color.

Chemiluminescent Detection with Lumigen PPD or CSPD

Note: This kind of detection can be only performed by using nylon membranes. Nitrocellulose causes a drastic decrease in sensitivity (Boehringer Mannheim data sheet for CSPD).

1. Transfer the membrane into a new tube and wash briefly with DEPC-treated H_2O and thereafter with of 20 ml buffer A_Tween for 1 min.

2. Incubate with 20 ml of buffer B for 45 min.
3. Dilute anti-DIG conjugate 1:10000 to a final concentration of 75 mU/ml in buffer B.
4. Pour off the solution and incubate with 15 ml of freshly prepared antibody-solution for 30 min.
5. Transfer the membrane to a new tube to avoid an increasing of background caused by antibodies bound to tube wall.
6. Wash 2×15 min with 20 ml of buffer A_{Tween} to remove unbound antibody.
7. Discard the solution and equilibrate with 20 ml of buffer C for 2–5 min.
8. Prepare 5–10 ml substrate solution; dilute Lumigen PPD or CSPD stock solution 1:100 in buffer C (sterile diluted substrate can be stored at 4 °C protected from light and may be reused for up to two times).
9. Put the membrane into a sterile dish and distribute the substrate solution over the surface.
10. Incubate the membrane at room temperature for 5 min. The incubation could be performed alternatively between two polyethylene sheets. Place the membrane between the sheets and add approximately 0.5 ml of substrate solution per 100 cm² of membrane onto the surface of membrane. Make sure that all air bubbles are removed between the top sheet and the membrane. Perform the incubation.
11. Let excess liquid drip off the membrane, lay the back surface down on a sheet of 3 MM paper for a few seconds but do not allow to dry completely.
12. Seal the damp membrane in a plastic bag avoiding air bubbles.
13. Preincubate at 37 °C for 5–15 min.
14. Expose for 15–60 min to X-ray film at room temperature. The time of exposure can be further increased depending on the strength of specific signals and the background. Luminescence continues for at least 24 h, whereby multiple exposures may be taken. The signal intensity increase during the first hours.

Trouble shooting:

For quantitative Northern blot analysis chemiluminescent detection is recommended due to its higher sensitivity. It has been stated that, for colorimetric systems, the activity of alkaline phosphatase is inhibited as the color precipitate is formed and deposited on filter. This enzyme inhibition would result in unreliable data [18].

If the sensitivity is too low:

- Estimate the labeling efficiency according to the protocol in section 2.6.2.4. Use longer PCR products for generation of probes.
- Increase concentration of probe during hybridization and/or prolong the hybridization time.
- Increase the antibody conjugate concentration and/or the substrate reaction time.

- By application of chemiluminescent detection, increase the time of pre-incubation at 37 °C and/or the exposure to X-ray film.

If the background is to high:

- Check the purity of the DNA fragments used for labeling reaction.
- Completely remove the formaldehyde from the gel after electrophoresis.
- Increase the concentration of the blocking reagent up to 5% for both the hybridization and immunological detection step.
- Take care that membranes do not overlap during washing.

To avoid both pitfalls, optimize the washing conditions (time, temperature, content of buffers) for your particular problem.

2.6.6 Analysis of Northern Blots

2.6.6.1 Evaluation of mRNA Size

Calculate the size of a detected mRNA species against a in parallel run RNA molecular weight marker (external standard) or against the 18 and 28 S rRNA species of the sample (internal standard).

Measure the distance from the loading well to each of the bands on the photographed gel.

Plot $\log_{10}$ of marker size against the distance migrated. Use the resulting curve to calculate the size of the detected RNA species.

Trouble Shooting:

Ethidium bromide retards the electrophoretic mobility of nucleic acids by about 15% [4].

If the molecular weight marker used, in contrast to the sample, is electrophoretically separated in presence of this dye, this fact must be considered for the calculation of molecular weights.

We recommend using DIG-labeled RNA molecular weight markers (Boehringer, Mannheim). These markers become visible automatically during detection and enables a direct size-calculation of a specific RNA on membrane. They may also serve as a check for successful transfer.

2.6.6.2 Semiquantitative Evaluation of Steady-State Levels of Specific mRNA

Target RNA expression levels can be measured following scanning densitometry of chemiluminescent or colorimetric detected DNA-RNA-hybrids.

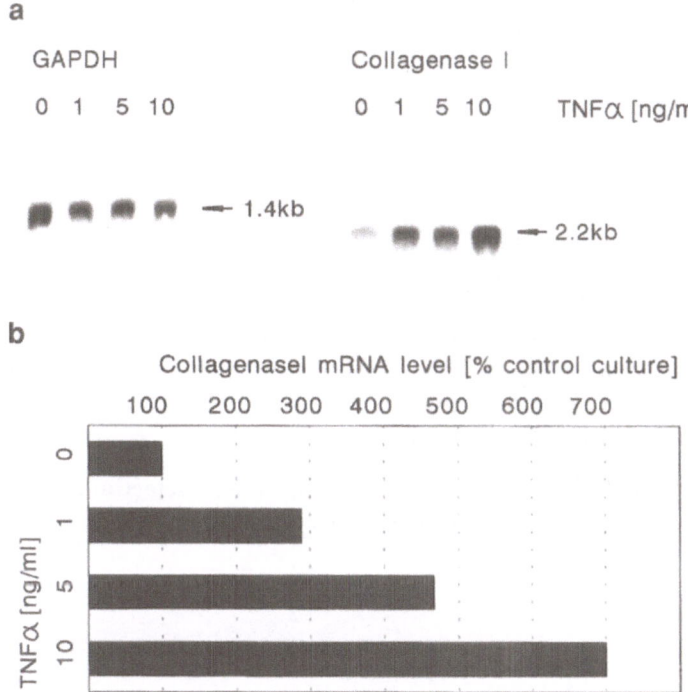

Fig. 2.6-4. Nonradioactive Northern blot hybridization with DIG-labeled dsDNA probes gene-rated by PCR: **a** Human Collagenase I and GAPDH specific probes were prepared according to the method described above (see section 2.6.2.). cDNA from human skin fibroblasts RNA was applied as template for amplification. For Collagenase I, sense primer was: 5′ AGGTATTG AGGGGAT GCT 3′ (635–653 bp) and antisense primer was 5′ GGTAGAAGGGATTTGTGCG 3′ (978–996 bp) [21], predicted PCR product size: 362 bp; for GAPDH primers see chapter 1.2, predicted PCR product size: 300 bp.
Human skin fibroblasts were incubated with various concentrations of human recombinant TNFα (1–10 ng/ml culture) for 24 h. 13 μg total RNA was analyzed by Northern blot hybridiza-tion with DNA probe specific for Collagenase I. To normalize the values obtained after hybridization the relative amount of the housekeeping gene GAPDH was determined parallel in each RNA sample. RNA-DNA hybrids were colorimetrically detected (duration of incubation: 16 h). **b** Relative Collagenase I mRNA levels were determined by scanning densitometry using the Epson GT-6000 scanner in combination with Gelimage software, Vers. 1.3 (Pharmacia) after colorimetric detection. The values were corrected for GAPDH mRNA levels in the same RNA preparations and expressed in relation to control culture incubated in parallel without TNFα

To normalize the densitometric values the simultaneous hybridization of each sample with a "housekeeping gene" specific probe (see chapter 1.1.) is necessary.

The application of the described nonradioactive Northern blot hybridiza-tion can be demonstrated ideally by the determination of Collagenase-I-mRNA-steady-state-levels of human skin fibroblasts incubated with recombinant TNFα (Figure 2.6-4).

2.6.7 Stripping and Reprobing of Northern Blots After Nonradioactive Detection

The application of nylon membranes for Northern blot hybridization enables the removal of hybridized probes following detection by extremely rigorous washing.

Thus one membrane can be screened for expression of various mRNA species using different probes of interest. This technique is very useful because the preparation of Northern blots is time-consuming and relatively large amounts of RNA are needed.

From colorimetrically stained membranes the color precipitate must be removed first followed by removal the probe. A less intense treatment is required after chemiluminescent detection.

Membranes can be reprobed up to ten times.

>>The membrane must keep wet for reprobing after detection! It is significantly more difficult to remove the hybridized probe from membranes, that have been allowed to dry completely.<<

Materials:

1) *Probe-stripping solution: 50% (v/v) formamide; 1% (w/v) SDS; 50 mmol/l Tris/HCl; pH 8.0 (20 °C); prepare the solution with DEPC-treated H_2O. Store the solution at 4 °C. Use about 20 ml of solution per 100 cm^2 of membrane.*
2) *$2 \times SSC$; prepared form a $20 \times SSC$ stock solution by using DEPC-treated H_2O (see appendix)*
3) *DEPC-treated H_2O*
   In addition DMFA is required for stripping membranes after colorimetric detection.

Protocol 8: Reprobing of Northern Blots

All the following incubations require shaking or mixing.

Stripping Membrane After Colorimetric Detection

1. Warm up about 100 ml DMFA to 50 – 60 °C in a water bath under a hood.

>>DMFA is volatile and can ignited above 67 °C!<<

2. Removal of color precipitate:
    Incubate membrane in prewarmed DMFA at 50 – 60 °C until the blue color precipitate is removed completely. Frequently changing the DMFA solution will increase the speed of reaction.
3. Transfer the membrane to a new plastic tube, and wash thoroughly with DEPC-treated H_2O at room temperature.

4. Removal of probe using a modified method according to Gebeyhu et al. [20]: Incubate the membrane for 2×30 min in about 20 ml prewarmed probe-stripping solution at 68 °C. Discard the solution, and wash the membrane with DEPC-treated H_2O and thereafter twice with $2 \times SSC$.

Stripping Membrane After Chemiluminescent Detection

1. Rinse the membrane thoroughly with DEPC-treated H_2O at room temperature.
2. Remove the probe as described above (see step 4).
   The membranes can now be used for prehybridization and hybridization with another probe or can be stored in $2 \times SSC$ in a sealed plastic bag for later applications.

References

1. Alwine JC, Kemp DJ, Stark GR (1977) Method for detection of specific RNAs in agarose gels by transfer to diazobenzyloxymethyl-paper and hybridization with DNA probes. Proc Natl Acad Sci USA; 74:5350
2. Kessler C, Höltke HJ, Seibl R, Burg J, Mühlegger J (1990) Nonradioactive labeling and detection of nucleic acids: I. A novel DNA labeling and detection system based on digoxigenin:anti-digoxigenin ELISA principle (digoxigenin system). Mol Gen Hoppe-Seyler; 371:917–927
3. Höltke HJ, Seibl R, Burg J, Mühlegger J, Kessler C (1990) Nonradioactive labeling and detection of nucleic acids: II. Optimization of the digoxigenin system. Mol Gen Hoppe-Seyler; 371:929–938
4. Farrel RE Jr (1993) RNA Methodologies. Academic Press, San Diego, New York, Boston, London, Sydney, Tokyo, Toronto
5. Engler-Blum G, Meier M, Frank J, Müller G (1992) Reduction of background problems in nonradioactive Northern an d Southern hybridization blot analysis enables higher sensitivity than ^{32}P-based hybridization. Anal Biochem; 210:235–244
6. Höltke H-J, Kessler Ch (1990) Non-radioactive labeling of RNA transcripts in vitro with the hapten digoxigenin (DIG); hybridization and ELISA-based detection. Nucl Acids Res; 18: 5843–5851
7. Kessler C (ed) (1992) Nonradioactive labeling and detection of biomolecules. Springer, Berlin, Heidelberg, New York, pp. 206–211
8. Yamaguchi K, Zhang D, Byrn RA (1994) Modified nonradioactive method for Northern blot analysis. Anal Biochem; 218:343–346
9. Sato M, Murao K, Mizobuchi M Takahara J (1993) Quantitative and sensitive Northern blot hybridization using PCR-generated DNA probes labeled with Digoxigenin by nick translation. BioTechniques; 15:880–882
11. Sambrook J, Fritsch EF, Maniatis T (1989) Molecular cloning: A Laboratory Manual. Cold Spring Harbor Laboratory Press New York. second edition
12. Feinberg AP, Vogelstein B (1983) A technique for radiolabeling DNA restriction endonuclease fragments to high specificity. Anal Biochem; 132:6–13
13. Rave N, Crkvenjakov R, Boedtker H (1979) Identification of procollagen mRNAs transferred to diazobenzyloxymethyl paper from formaldehyde gels. Nucl Acids Res; 6:3559
14. Lehrbach H, Diamond D, Wozney JM, Boedtker H (1977) RNA molecular weight determinations by electrophoresis under denaturing conditions critical reexamination. Biochemistry; 16:4743
15. Mc Master GK, Carmichael GC (1977) Analysis of single- and double-stranded nucleic acids on polyacrylamide and agarose gels and acridine orange. Proc Natl Acad Sci U.S.A., 74: 835

16. Bailey JM, Davidson N (1976) Methylmercury as reversible denaturing agent for agarose gel electrophoresis. Anal Biochem; 70:75
17. Church GM, Gilbert W (1984) Genomic sequencing. Proc Natl Acad Sci U.S.A.; 81:1991–1995
18. Pitta A, Considine K, Braman J (1990) FLASH chemiluminescent system for sensitive non-radioactive detection of nucleic acids. Strategies in Molecular Biology; 3:33
19. Goldberg GI, Wilhelm SM, Kronberger A, Bauer EA, Grant GA, Eisen AZ (1986) Human fibroblast collagenase: complete primary structure and homology to an oncogene transformation induced rat protein. J Biol Chem; 261:6600–6605
20. Gebeyehu G, Rao PY, SooChan P, Simms DA, Klevan L (1987) Novel biotinylated nucleotide-analogs for labeling and colorimetric detection of DNA. Nucl Acids Res; 15:4531–4534

Part 3

**Semiquantitative and Quantitative Protocols
for Measurement
of Nucleic Acids by PCR**

Chapter 3.1

Quantitation of mRNA by the ELOSA Technique Using External Standards

D. LASSNER

3.1.1 Principle of the ELOSA Technique

In most studies, PCR applications have been coupled with either a gel electrophoresis step or liquid/solid support hybridization (see chapter 2.5). Many detection methods are very time consuming or their sensitivity is too low [1].

The ELOSA (Enzyme Linked Oligonucleotide Sorbents Assay) combines both sensitivity and specificity of a probe hybridization technique with rapid and simple methods of detecting and quantifying PCR products in various microplate formats [2]. This assay is mainly comparable to the ELISA technique which is an essential detection method in clinical diagnostics. The specific recognition step is performed by hybridization of amplified DNA fragments with a probe.

PCR could realize its full potential in clinical laboratory with ELOSA. This technique increases the throughput of clinical samples with rapid turnover. ELOSA only needs commercial available chemicals and does not require radioactive isotopes to increase sensitivity [4]. Results can be achieved in less than 4 h, whereas conventional hybridization methods with comparable specificity and sensitivity need up to three days [2].

ELOSA is a four-step method:

1) Immobilization of a probe or PCR-amplified DNA fragment in a microwell
2) Denaturation of double stranded DNA to get two single-strands
3) Hybridization of probe to the complementary single-strand
4) Colorimetric detection of hybrids (probe/DNA)

The combination of polymerase chain reaction and detection of achieved products by ELOSA is called PCR-ELOSA [2, 3]. The specificity of PCR-ELOSA is given by an amplification with gene specific primers followed by a hybridization step using a probe which detects intrinsically generated PCR products. Sensitivity depends on exponential accumulation of newly synthesized DNA.

The procedure of detection is shown schematically in Figure 3.1-1.

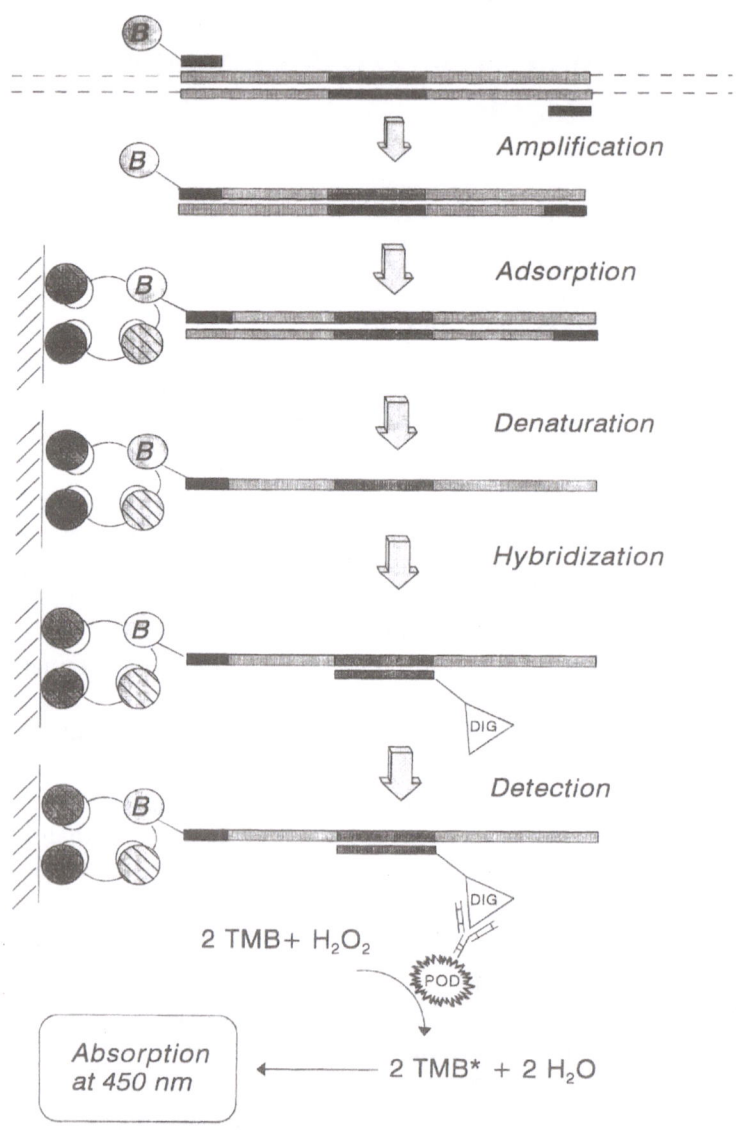

Fig. 3.1-1. Principle of PCR-ELOSA: PCR product is generated by amplification with specific primers (one is biotinylated) and coupled to a streptavidin-coated microtiter plate. Single-stranded DNA fragments are achieved by denaturation with NaOH. A labeled probe is hybridized to the complementary sequence followed by colorimetrically detection of the DNA/probe hybrids by using anti-DIG-antibody-POD-conjugates and subsequent reaction with chromogen TMB

The linkage of two very sensitive and variable detection methods increase the applications of this technique in research and clinical diagnostics. PCR-ELOSA is suitable for qualitative detection of amplification products or for quantitation of DNA [4] or mRNA [3]

3.1.2 Quantitation of mRNAs by PCR-ELOSA with External RNA Standards

Competitive RT-PCR using internal standards (see chapter 3.4) is the most sensitive and accurate method of mRNA quantitation [5]. Most of the described quantitative protocols are based on simple comparison of the amount of specific mRNA versus a reference nucleic acid amplified by the same way. The results achieved may be indicated as "arbitrary units" [6].

A similar approach is the parallel amplification of serial dilutions of definite amounts of RNA standards and the unknown sample followed by detection of synthesized PCR products by ELOSA. RNA standards can be designed as described in chapter 1.2 and 1.3.

DNA fragments detected by the ELOSA technique are prepared by previously performed PCR amplification. Specificity of the PCR process strongly influences the quantitation performed by the subsequent hybridization step. Simultaneous processing of multiple samples ensures comparable results. The microplate format of detection offers the possibility of fast and reliable analysis of up to 96 samples to get numeric data [3].

In Table 3.1-1 advantages and disadvantages of the ELOSA assay and the well established HPLC method [7] are compared.

Table 3.1-1. Comparison of PCR-ELOSA and HPLC quantitation of PCR products

Method	HPLC	ELOSA
Sensitivity	200 – 400 pg	10 – 20 pg
Detection of	amount and size	amount and specific sequence
Accuracy	± 4 %	± 10 %
Turnover of samples: analysis	15 – 30 min /sample	1 – 5 h/96 samples
automatization possible	Yes	Yes

3.1.3 Experimental Procedures

The amounts of accumulated product from amplifications with Taq or rTth DNA polymerase (chapter 2.4) are determined by the ELOSA technique. Knowing the initial concentration of mdr1-RNA (cRNA standards) and the absorbance after colorimetric detection of the PCR product, the sample mdr1-mRNA content can be calculated.

The cRNA standards should be as handled as the samples containing unknown amounts of the mRNA of interest. The RT reaction should be performed by specific priming (see chapter 2.2) and the whole sample should be used in following single-tube RT-PCR. A prerequisite of detection is the use of a biotinylated upstream primer for amplification (chapter 2.4).

All measurements should be performed in the exponential phase of PCR amplification. Therefore, the number of cycles representing the linear phase of reaction should be known before starting the experiment (see chapters 1.1 and 1.2).

3.1.3.1 Quantitation of mdr-1 mRNA by PCR-ELOSA

RT-PCR is performed as described in chapters 2.2 and 2.4 using parallel amplified standard dilutions of definite amounts of mdr-1 cRNA for calibration of the assay (see chapter 1.3). The amplification kinetics of the cRNA standards may be followed starting with cycle 12. The plateau phase is reached at cycle 30. Therefore the amplification process is usually stopped between cycle 24 and 26 to get reliable results.

PCR samples are firstly roughly examined by agarose gel electrophoresis as described (see section 2.3.2.2). The samples are diluted 1:80 in the first step of examination by ELOSA. Samples that are only weakly amplified and therefore not visible in ethidium bromide stained gels are diluted 1:20 as mentioned below.

The first step of ELOSA is coupling of a biotinylated PCR product to a streptavidin-coated microwell plate. By addition of NaOH to the adsorbed dsDNA fragments the DNA strand containing the non-biotinylated primer is removed. The microplate adsorbed ssDNA can now be detected using a strand-complementary DIG labeled probe by colorimetric detection.

The following protocol has been established for simple handling. The use of microplate washer and pipettor increase reproducibility of the assay.

Materials:

1) *Phosphate-buffered saline (PBS): see appendix*
   - *Solution A: PBS with 0.1 % Tween 20*
   - *Solution B: PBS with 2.0 % Tween 20*
2) *Hybridization buffer 20 × SSPE (pH 7.0)*
   - *175.3 g NaCl*
   - *27.6 g $NaH_2PO_4 * H_2O$*
   - *7.4 g $EDTA-Na_2 * 2H_2O$*
   - *Aqua dest. add 1 l*
3) *Wash buffer (Phosphate-buffer; pH 6.4)*
   - *1.045 g $NaH_2PO_4 * H_2O$*
   - *2.40 g $Na_2HPO_4 * 12 H_2O$*
   - *7.60 g NaCl*
   - *Aqua dest. add 1 l*

4) *Detection solutions*
   - *Solution C:*
     *Prepare a solution of 8.9 g citric acid * H_2O in 160 ml aqua dest. Adjust pH value of 5.0 by addition of 4N KOH and fill up to 200 ml with aqua dest. Add 200 µl of H_2O_2 (30 %, v/v).*
   - *Solution D: TMB chromogen*
     Dissolve 4.8 mg TMB (Boehringer, Mannheim) in 780 µl DMFA
   - *Mix 1 vol. solution D + 9 vol. solution C immediately before use.*
5) *Denaturation solution: 0.25 mol/l NaOH*
6) *Stop solution G: 2N HCl*
7) *Anti-DIG-POD conjugate (Boehringer, Mannheim)*
8) *Streptavidin-coated microwell plate strips (DuPont)*
9) *DIG-labeled mdr-1.probe (0.4 ng/100 µl 0.5 SSPE-B.)*
10) *Combipette and combitips (Eppendorf)*
11) *Microplate reader (e.g. Anthos)*
12) *Microplate washer (e.g. Denley Wellwash)*
13) *Incubator (37 °C and 50 °C) with shaking table*

Protocol 1: Quantitation of the mdr-1 PCR Product by ELOSA

1. Dilute 10 µl of each PCR sample depending on the predicted DNA amount (1:20 or 1:80) in solution A.
2. Pipette 100 µl of dilution from each PCR sample to the respective well on the microwell plate strip and incubate at 37 °C for 60 min.
3. Remove all solution from the wells by rapid inversion of the strip holder and shaking down vigorously.
4. Wash the strips 3 times with 300 µl wash buffer (microplate washer).
5. Wash once with 200 µl solution A.
6. Pipette 100 µl denaturation solution to each well and incubate for 10 min at room temperature.

>>**Do not exceed this time**<<

7. Wash immediately 3 times with 300 µl wash buffer (microplate washer).
8. Pipette 200 µl solution B and incubate for 15 min at room temperature.
9. Wash 6 times with 300 µl wash buffer (microplate washer).
10. Pipette 100 µl DIG-labeled probe to each well and incubate at 50 °C for 1 h.
11. Wash 6 times with 300 µl wash buffer (microplate washer).
12. Add 100 µl anti-DIG-POD conjugate to each well and incubate at 37 °C for 15 min.
13. Prepare detection solution by mixing solution C and solution D.
14. Wash all wells 6 times with 300 µl wash buffer (microplate washer).
15. Pipette 100 µl detection solution to each well, shield the plate from light and incubate at 37 °C for 15 min on a shaking table.
16. Stop reaction by adding 100 µl stop solution.
17. Read the absorbance of each well at A_{450}.

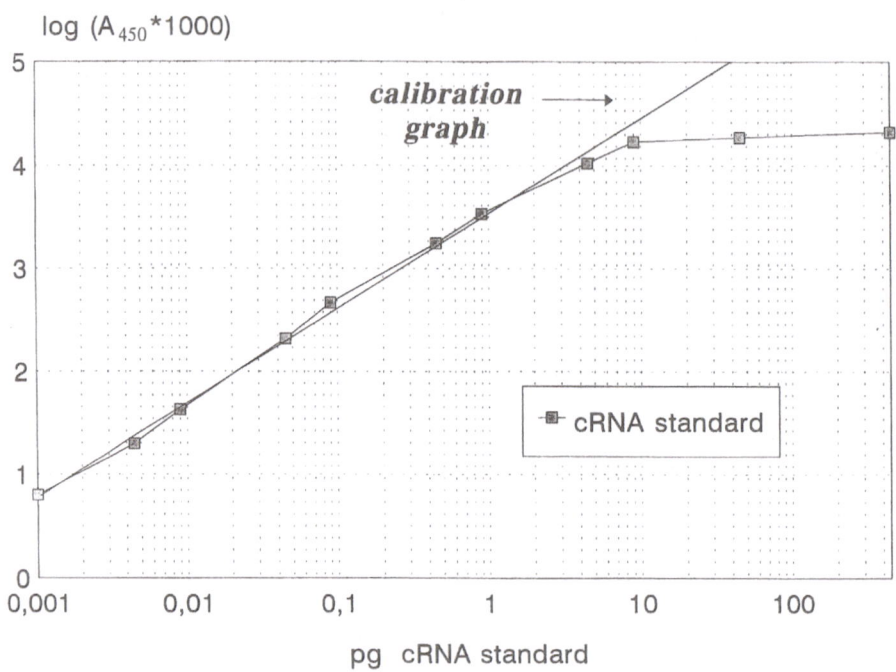

Fig. 3.1-2. Plot of measured absorbance versus cRNA content to obtain the calibration graph

Water blanks and PCR blanks (samples without RNA) should be treated simultaneously. Calculate the results by subtracting the blank from the corresponding sample values. Plot the logarithm of absorbances obtained from the standard dilution series versus Log of the cRNA amount introduced into the PCR mixture. Analyze the exponential phase of amplification by linear regression and use the resulting graph for calibration of results derived from the analyzed sample. Calculate sample mdr-1 mRNA content by comparison of absorbance in the logarithmic form with the calibration graph as shown in Figure 3.1-2.

Trouble Shooting:

Using microplate assays, it is necessary to remove all solutions needed for enzymatic reaction or washing steps as completely as possible. The most critical step is incubation with antibody-conjugate which, when nonspecifically adsorbed to the well, strongly influences the blank absorbance. But in general, increased background signals can be eliminated by subtraction of blank absorbance from sample absorbance.

Do not exceed the time necessary for the denaturation step with NaOH. Excessively long alkaline treatment leads to partial loss of immobilized ssDNA from the microplate. Be sure that the biotin-linkage of the primer is stable under the selected conditions.

3.1.3.2 Sensitivity and Reproducibility of the Assay

The excellent reproducibility of the assay is demonstrated by an interassay coefficient of about 15% for detection of mdr-1 mRNA. The intraassay coefficient was shown to be below 12%. These results correspond well with comparable data achieved with commercially available ELOSA kits [4, 8].

When stored at −20 °C the cRNA dilution series were stable for at least two months, indicated by obtaining nearly identical calibration curves. In contrast, solutions used for RT-PCR and for ELOSA detection should be freshly prepared.

References

1. White TJ, Madej R, Persing DH (1992) The polymerase chain reaction: clinical application. Adv Clin Chem; 29:161–196
2. Adler K, Buzby P, Bobrow M, Erickson T (1992) Rapid DNA detection using a fluorescein-antifluorescein-based (FAF) ELOSA. DuPont Biotech Update; 7:13–15
3. Alard P, Lantz O, Sebagh M, Calvo CF, Weill D, Chavanel G, Senik A, Charpentier B (1993) A versatile ELISA-PCR assay for mRNA quantitation from a few cells. Biotechniques; 15:730–737
4. Jalava T, Lehtovaara P, Kallio A, Ranki M, Söderlund H (1993) Quantification of Hepatitis B virus DNA by competitive amplification and hybridization on microplates. Biotechniques; 15:134–139
5. De Kant E, Rochlitz CF, Herrmann R (1994) Gene expression analysis by a competitive and differential PCR with antisense competitors. Biotechniques; 17:934–942
6. Burger H, Nooter K, Zaman GJR, Sonneveld P, van Wingerden KE, Oostrum RG, Stoter G (1994) Expression of the multidrug resistance-associated protein (MRP) in acute and chronic leukemias. Leukemia; 8:990–997
7. Katz ED, Dong MW (1990) Rapid analysis and purification of polymerase chain reaction products by high-performance liquid chromatography. Biotechniques; 8:546–555
8. Handbook Amplicor "*Chlamydia trachomatis* test", Hoffmann-La Roche 1993

Chapter 3.2

Semiquantitative Detection of Viral DNA, e. g. for CMV, by Using the DNA Enzyme Immunoassay (DEIA)

B. Pustowoit

3.2.1 CMV Infection in Human Beings

Cytomegalovirus, consisting of a unique double-stranded DNA genome 240 kb in length, is a member of the Herpes virus family. Herpes viruses are ubiquitous agents which infect humans and can cause high morbidity and mortality [1].

Specifically, productive CMV infection has been associated with retinopathy, hepatitis, gastroenteritis, pneumonia and 10% of infectious mononucleosis cases. About 0.5 to 2.5% of infants are congenitally infected. The clinical manifestation of CMV infection is a significant problem in cases of immuno-compromised individuals such as AIDS patients and recipients of organ and bone-marrow transplants. CMV pneumonitis is a serious complication in transplant recipients and patients with AIDS. CMV infects an estimated 95% of all AIDS patients. Early detection of CMV is crucial since patients may now be treated with ganciclovir [19].

3.2.2 Applications of Quantitative PCR in Cytomegalovirus Pathogenesis

The concept of modulations in viral load during reactivation of a latent CMV infection can be envisaged as a series of peaks and troughs of viral load over time.

Two extremes of assay for CMV can be envisaged; an assay that will not detect the virus in the clinical sample (i. e. less sensitive that the peak of the virus load) and an assay that is so sensitive that it detects the virus at all stages of infections (i. e. sensitive enough to detect latent virus loads).

However, the load is limited by the ultimate level of sensitivity. This can be an application of quantitative PCR methods to view modulation in cytomegalo-viral load.

3.2.3 Definition of CMV Infection

1. Viremia being CMV detected from blood or blood cells by culture.
2. Antigenemia being CMV antigen detected by monoclonal antibodies from leukocytes.

3. PCR based techniques should define the clinical specimen as well as the type of PCR used.
4. CMV disease should require symptoms and/or signs from the affected organ together with CMV detected from that organ. In many situations, in particular in HIV patients, the absence of other explanations is also a requirement [20].

3.2.4 Diagnosis of Cytomegalovirus Infection

There are numerous methods and approaches to diagnose CMV infection in patients and normal individuals. Each of them has its advantages and disadvantages. The optimal method frequently depends on the nature of the clinical samples collected and the clinical manifestation at the time a specimen is collected.

Therefore other measurements such as detecting CMV in urine by PCR or culture isolation is justified.

In Figure 3.2-1 the cytopathic effect of cytomegalovirus on humane fibroblasts is demonstrated.

The virus has changed the morphology of the cells so that they have become spherical. Such a cytopathic effect is really seen only after 3 or 5 days of cultivation. The result of this cultivation shows very clearly that in the patient material there was enough virus for growing in cell culture. The test is the so called "golden standard" for CMV diagnosis.

The superb sensitivity and specificity of PCR makes it one of the best methods in the clinical laboratory for detection in a great variety of clinical samples. Using this method, very small quantities of viral DNA are detectable.

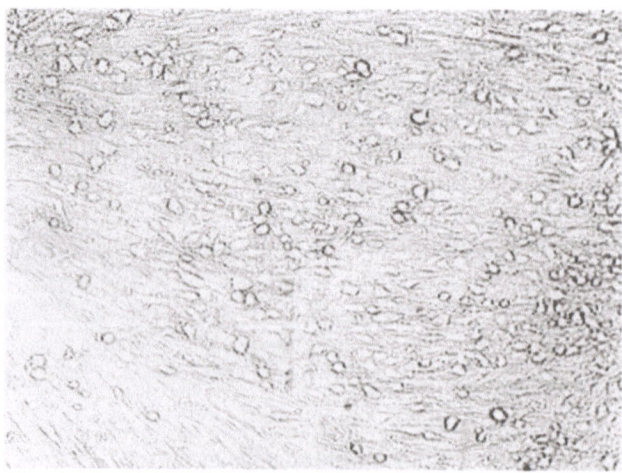

Fig. 3.2-1. Typically cytopathic effect (CPE) from CMV in human lung fibroblasts

One of the seriously complication of CMV infection is interstitial pneumonia. In this case Bronchial alveolar lavage fluid (BAL) must be investigated.

PCR has been found to be the most sensitive method for detection of CMV in BAL fluid. A result was obtained within 5 h and no processing of samples was required. Ericsson et al. [2] showed, that from 52 investigated clinical samples, 15 patients proved positive by conventional virus isolation and further 7 patients were positive by PCR only. The authors believe, that a negative CMV PCR has a high negative predictive value. None of the 30 patients CMV negative by PCR developed CMV pneumonia within the following 2 months.

The study focuses on the problem of evaluation the specificity of a positive PCR test when concomitant virus isolation is negative. This discrepancy between the methods might be explained either by contamination or by detection of incomplete or non-viable virus particles by the PCR method. Control tests did not indicate false-positive tests when we ourselves used the PCR method for CMV. Loss of infectivity because of unfavorable conditions during transportation or storage cannot be excluded, although we did optimize the handling of the samples.

Prösch et al. [3] described the monitoring of patients by centrifugation and PCR for cytomegalovirus after organ transplantation. PCR was more sensitive in an earlier phase of infection when the virus detection in the cell culture was still negative.

PCR technology has also been involved in the prenatal diagnosis of pregnancies which are at risk for congenital cytomegalovirus infection [4]. In an intrauterine CMV infection, the fetus would excrete CMV via urine into the amniotic fluid. The amniotic fluid therefore seems a logical choice of bodily fluid for the prenatal diagnosis of CMV transmission [5]. The virus can be found by culture or PCR in this material. Fetal blood taken by cordocentesis could represent an alternative pathological material. At present there is not enough evidence to predict fetal outcome, therefore the question "Is the fetal infection going to be damaging?" remains unanswered.

Each year in the United States, approximately 40 000 infants are infected congenitally with cytomegalovirus [6]. Of these, 10 % are born symptomatic, and up to 95 % will have neurologic sequelae. CMV is rarely isolated from the cerebrospinal fluid (CSF) [7]. Atkins et al. [8] used the PCR on CSF from infants with congenital infection with CMV to determine if CMV DNA could be amplified and detected, and they correlated its presence with CNS disease.

The CMV show tropism to very different organs with different clinically pictures in the patients. One of the complications often found after bone marrow transplantation is enteritis/colitis. In this case the patient material must be taken using an invasive method. Some investigators are now trying to determine the CMV-DNA directly in stool. The results from Michel et al. [9] indicate that a lack of evidence from HCMV-DNA in stool samples excluded a gastrointestinal infection in the patient and a positive result is a sign of a HCMV-infection of the ileum.

PCR in aqueous humor is currently the only methodological approach to an etiological diagnosis of HCMV retinitis, PCR results are consistently conclusive and can be reliably used for the local monitoring of the efficacy of antiviral treatment.

A highly efficient single cell PCR method has been developed, using a FACS-based automated cell disposition unit, which sorts single cells directly into PCR reaction tubes. The test system allows the determination of the percentage of cell-specific viral DNA [10].

3.2.5 Experimental Procedures

3.2.5.1 Preparation of Samples for PCR – Principal Difficulties

The PCR technique can be used to detect both virus-specific DNA or mRNA. For viral-RNA detection, an additional reverse transcription step is needed to generate either single-stranded or double-stranded DNA for a template. This involves the addition of oligo (dT) and reverse transcriptase to synthesize complementary DNA for subsequent amplification [11].

Rapid and efficient purification procedures are necessary in order to guarantee high efficiency in subsequent enzymatic reactions.

It was shown that which clinically symptoms caused by a CMV infection determined which patient material was taken.

Total DNA for PCR can be purified from different cells, body fluids, peripheral blood cells and stool. The consistence of the patient material is very important for the preparation of patient materials. Peripheral blood cells are the most common clinical samples encountered for CMV detection.

This is the reason for our presenting quantitative PCR here as the technique for this patient material.

The preparation (here described for EDTA or heparin blood) was taken from Boom et al. [12].

Materials:

1) *Lysis buffer for the erythrocytes:*
   60 g sucrose + 1 g $MgCl_2 \times 6H_2O$ + 0.6 g Tris + 5 ml Triton X 100, add water until 400 ml is reached, bring the pH to 7.5 using HCl and add the rest of water until 500 ml
2) *buffer L2:*
   120 g Guanidinthiocyanate (GuSCN) in 100 ml Tris/HCl pH 6.4
3) *buffer L6:*
   buffer L2 + 22 ml 0.2 M EDTA pH 8.0 + 2.6 g Triton X 100

Protocol 1: Sample Preparation from Blood

1. Add 900 µl erythrocyte lysis buffer to 200 µl of the blood sample from the patient.

2. Mix and centrifuge for 30 s at high speed.
3. Wash the pellet twice with 900 μl lysis buffer L2.
4. Add 900 μl lysis buffer L6.
5. Add 100 μl fractionated silica.
6. Mix, wait for 10 min at 21 °C, mix again, centrifuge for 10 s.
7. Wash the pellet twice with ethanol 70 % and once with acetone.
8. Dry for 10 min at 56 °C (vacuum).
9. Add 100 μl of 10 mmol/l Tris/HCl, 1 mmol/l EDTA (pH 8.0).
10. Mix, incubate for 10 min at 56 °C.
11. Mix, centrifuge (2 min, 13 000 rpm).
12. Take the supernatant, clear by short centrifuging and use aliquots of the supernatant for PCR.

>>**Washing procedure: it is important that the silica pellet is completely resuspended in the washing solutions. In the first wash, the pellet can be very tight and it may take up to 1 min to resuspend such pellets by vortexing.**<<

3.2.5.2 Detection of Human Cytomegalovirus Using PCR

Materials:

1) Primers:
   The oligonucleotide primers were synthesized with an Applied Biosystems DNA Synthesizer. The sequences were (5′ to 3′): AGACCTTCATGCAGATCC (up-stream) and GGTGCT CACGCACATTGATC (down-stream) for the primer set complementary to the IE1 gene [13,14].
2) Reagents and equipment necessary to perform a standard PCR reaction (see chapter 2.3.)

Protocol 2: CMV-PCR

- Mastermix (calculated for 10 samples, prepared cold):
  - 330 μl ddH$_2$0
  - 50 μl Taq buffer (10 x)
  - 50 μl dNTP's (1 mmol/l)
  - 20 μl of each described IE1-primer
  - 3 μl Taq polymerase (5 U/μl)
- Add 5 μl of prepared patient sample or controls to 45 μl of the mastermix.
- Place the cold PCR samples into the 95 °C preheated thermocycler (GeneAmp 9600) and amplify by the following protocol:

1. linearise and denature	2. amplify 40 cycles	3. final extension	4. store till further use
3′ – 95 °C	30″ – 55 °C 30″ – 72 °C 20″ – 94 °C	15′ – 72 °C	endless 4 °C

- After PCR the samples have been denaturated for 15 min at 95 °C, they can be used for the DEIA.

3.2.5.3 DEIA: DNA Enzyme Immunoassay

We recommend using the technique as reported by Mantero et al. [15]. The DNA Enzyme Immunoassay (DEIA) can be used to reveal specific DNA sequences over a wide range of concentrations and molecular sizes. It is particularly suitable in the specific detection of amplified DNA by any documented technique.

Principle of the Assay

The DNA Enzyme Immunoassay is based on the hybridization of amplified DNA with an oligonucleotid probe, coated on the wall of the microtiter plate wells with an avidin-biotin bond. The probe is designed to identify a region of the IE1 gene, which codes the major immediate early antigen of HCMV [16]. This gene region has been selected from the preserved areas between different strains and viral isolates.

The hybrid of the probe and the DNA being examined is revealed in an original way by using an anti-DNA monoclonal antibody. This antibody only reacts with double-stranded DNA and not with the single-stranded DNA. When the denatured PCR products are dispensed into the wells, the probe binds to the complementary sequence in positive samples; on the other hand if the sample is negative, it will not contain the sequence under examination, and the hybrid (double-strand molecular species) will not form.

After incubation and subsequent clearing of the sample, the addition of the double stranded anti-DNA antibody identifies the wells where hybridization has taken place (antibody binding, positive reaction) and the wells where the probe has not bound the DNA which has remained in the single strand form (no antibody binding, negative reaction). The addition of an enzyme tracer (anti mouse IgG labeled with peroxidase) will reveal the DNA/antibody bond.

Test Procedure

For the determination the test kit GEN-ETI-K HCMV, Sorin Biomedica, Italy was used.

Protocol 3: Detection of Amplified CMV DNA Fragments by DEIA

1. Dispense 100 µl hybridization buffer in each well.
2. Dispense 20 µl negative control, positive controls and denatured samples.
3. Incubate for 1 h at 50 °C.
4. At the end of the first incubation, prepare the Anti-DS-DNA solution.
5. Wash the strips 3 times with washing buffer.
6. Dispense 100 µl diluted Anti-DS-DNA solution in all the wells except for the blank.
7. Incubate for 1 h at room temperature.
8. Prepare the enzyme tracer solution at the end of the second incubation.
9. Wash the strips 3 times with washing buffer.
10. Dispense 100 µl diluted enzyme tracer solution into all the wells except for the blank.
11. Incubate for 1 h at room temperature.
12. Prepare the chromogene/substrate at the end of the third incubation.
13. Wash the strips 3 times with washing buffer.
14. Dispense 100 µl chromogene/substrate into all the wells.
15. Incubate for 30 min at room temperature in the dark.
16. Dispense 200 µl stop solution into all the wells.

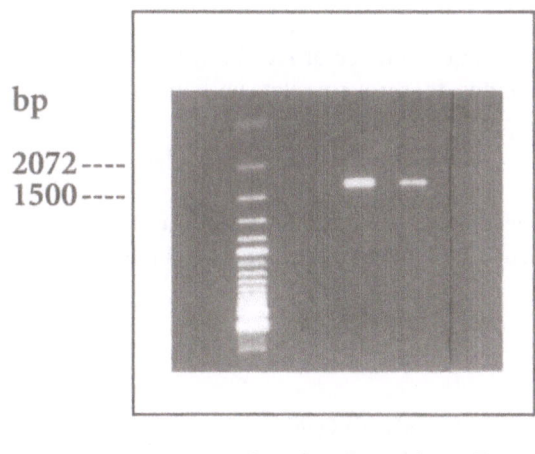

Fig. 3.2-2. Agarose gel electrophoresis of PCR amplified CMV DNA fragments. *1* – 100 bp DNA Ladder, *2* – cell culture without CMV (negative control), *3* – CMV-infected cell culture, un-diluted, *4* – CMV-infected cell culture, 1:10 diluted, *5* – CMV-infected cell culture 1:100 diluted; *last row:* extinction values obtained by DEIA assay

Calculation of Results

Reset the instrument with the blank, read the O.D. at 450 nm (reference wavelength is 630 nm). If an ETI- SYTEM READER with ETI-SYSTEM SOFTWARE (Sorin) is used, all the calculations are performed automatically.

Interpretation

The absorbance values obtained for the samples should be compared with an analytical cut-off value, which discriminates the positive specific hybridization from the negative one. From the evaluation of the analytical sensitivity and specificity, the suggested cut-off is calculated as average O.D. relative to the negative control plus 0.150 (Ncx + 0.150 O.D.). In Figure 3.2-2 typical patterns for agarose gel electrophoresis compared to O.D. for a dilution series are given.

3.2.6 Limitation of PCR for the Diagnosis of CMV Infection

All diagnostic tools in CMV diagnostics must be focused on finding out clinically significant results. Prösch et al. [17] showed that with CMV primary infections in patients with bone marrow transplantation, the HCMV-DNA-PCR from PBL was found to be positive one or two weeks earlier than cell culture. On the other hand, in latently infected patients, only after a delay of 2–5 weeks was an antigenemia noticeable. In this case, even the HCMV-DNA-PCR as the method giving the earliest results did not give positive results. The extreme sensitivity of PCR reaction can be achieved during identification of CMV in clinically asymptomatic patients. The evidence of HCMV-DNA from periphery mononuclear cells and granulocytes does not correlate with viremia specially in patients with ganciclovir-therapy.

On the other hand the standardization of HCMV-DNA-PCR is not yet well developed.

The best correlation between DNA-emia, virus isolation and illness was found with detection of the virus in the plasma of the patients [18]

References

1. De Marchi JM, Blankship GD, Brown GD, Kaplan AS (1994) Size and complexity of human cytomegalovirus DNA. Virology; 89:643–649
2. Eriksson B-M, Brytting M, Zweygberg-Wirgart B, Hillerdal G, Olding-Stenkvist E, Linde A (1993) Diagnosis of Cytomegalovirus in Bronchoalveolar Lavage by Polymerase Chain Reaction, in comparison with virus isolation and detection of viral antigen. Scand J Infect Dis; 25:421–427
3. Prösch S, Kimel V, Dawydowa I, Krüger D (1992) Monitoring of patients for cytomegalovirus after organ transplantation by centrifugation culture and PCR. J Med Virol; 38: 246–251

4. Donner C, Lienard C, Jean, Blain, Aderca, Juliette, Rodesch, Frederic (1993) Prenatal diagnosis of 52 pregnancies at risk for congenital cytomegalovirus infection. Obstetrics Gynecol; 82:481–486

5. Landini MP (1994) Recent anvances in cytomegalovirus infection and ist diagnosis. Abbott Symposium, Langen, 17 October 1994, oral presentation

6. Demmler GJ (1991) Infectious Diseases Society of America and Center for Disease Control: summary of a workshop on surveillance for congenital cytomegalovirus disease. Rev Infect Dis; 13:315–329

7. Jamison RM, Hathorn AW (1978) Isolation of cytomegalovirus from cerebrospinal fluid of a congenitally infected infant. Am J Dis Child; 132:63–64

8. Atkins JT, Demmler GJ, Williamson WD, McDonald JM, Allison SI, Buffone GJ (1994) Polymerase chain reaction to detect cytomegalovirus DNA in the cerebrospinal fluid of neonates with congenital infection. J Inf Disease; 169:1334–1337

9. Michel D, Marre E, Roczkos D, Mertens T. PCR aus Stuhlproben zum Ausschluß von Zytomegalievirus- verursachten Enteritiden bei immundefizienten Patienten. Frühjahrstagung der Gesellschaft für Virologie, Gießen 15.–18.3.1995

10. Bertram S, Hufert F, Neumann-Haefelin D, von Laer D. Detection of DNA in single cells using an automated cell deposition unit and PCR. Frühjahrstagung der Gesellschaft für Virologie, Gießen, 15.–18.3.1995

11. Huang E-S, Kowalik TF (1993) Diagnosis of human cytomegalovirus infection: laboratory approaches In: Becker Y, Darai G, Huang E-S (eds.). Molecular aspects of human cytomegalovirus diseases. Springer, Berlin Heidelberg New York, pp. 225–256

12. Boom R, Sol CJA, Salimans MMM, Jansen CL, Wertheim-Van Dillen P, van der Noordaa ..? (1990) Rapid and simple method for purification of nucleic acids. J Clin Microbiol; 28: 495–503

13. Stenberg RM, Thomsen DR, Stinski MF (1984) Structural analysis of the major immediate early gene of human cytomegalovirus. J Virol; 49:190–199

14. Akrigg A, Wilkinson G, Oram, JD (1985) The structure of the major immediate early gene of human cytomegalovirus strain AD 169. Virus Res; 2:107–127

15. Mantero G, Zonaro A, Albertini A, Bertolo P, Primi D (1991) DNA Enzyme Immunoassay: general method for detecting products of polymerase chain reaction. Clin Chem; 37:3

16. Zipeto A, Silini E, Parea M, Percivalle E, Zavattoni M, Di Matteo A, Milanesi G (1990) Identification of human cytomegalovirus isolated by the polymerase chain reaction. Microbiologica; 13:297–304

17. Prösch S, Schielke E, Meisel H, Einhäupl KM, Krüger DH. Diagnostischer Wert der HCMV PCR bei Knochenmarktransplantierten und Patienten mit HCMV Meningitis/Encephalitis. Frühjahrstagung der Gesellschaft für Virologie, Gießen 15.–18.3.1995

18. Hamprecht K, Sorg G, Steinmaßl M, Grebenstein A, Döller G, Jahn G. Diagnostische Wertigkeit des qualitativen HCMV-DNA-Nachweises aus peripheren mononukleären Zellen, Granulozyten und zellfreiem Plasma immunsupprimierter Patienten über nPCR. Frühjahrstagung der Gesellschaft für Virologie, Gießen 15.–18.3.1995

19. Emery VC (1993) Cytomegalovirus pathogenesis: Application of PCR. Michelson S, Plotkin SA (eds.) In: Multidisciplinary approach to understanding cytomegalovirus disease Excerpta Medica, International Congress Series 1032, Amsterdam, New York, London, Tokyo, pp. 63–75

20. Ljungman P, Griffiths P (1993) Definitions of cytomegalovirus infection and disease in multidisciplinary approach to understanding cytomegalovirus disease. Michelson S, Plotkin SA. Excerpta Medica International Series 1032, Amsterdam, New York, London, Tokyo, pp. 87–93

Chapter 3.3
HPLC – Analysis of Nucleic Acids

H. REMKE and TH. KÖHLER

3.3.1 Theoretical Background

High-performance-liquid-chromatography (HPLC) was developed for the detection of small organic compounds and proteins. Many different column materials have been developed for these analyses in relation to particle size, linker size or charge. These materials represent the stationary phase within a column responsible for retardation of substances contained in the samples according to their size and charge measurable by the specific retention times. The mobile phase is represented by the elution buffer which is pumped under high pressure through the stainless-steel-column.

Recently HPLC has also been employed for the purification and quantitation of nucleic acids, e.g. for plasmids [1] and PCR products [2, 3]. Nucleic acids can be analyzed by the "reversed phase" (RP) technique or by use of anion exchange columns [1–3].

A reliable HPLC method has been described for the purification and quantitation of double stranded DNA by use of very small non porous particles of anion exchangers. The DNA-fragments may be obtained by the polymerase chain reaction (PCR) or from enzymatic cleavage of nucleic acid duplexes [4]. In combination with RT/PCR the HPLC-method represents a reliable and powerful tool for quantitative studies on gene expression [3].

We use this method to analyze DNA fragments derived from competitive PCR-reactions. The main advantage is its ability to distinguish clearly between the main and side products and the calculation of the exact product ratios necessary for quantitation of absolute mRNA copy numbers [4].

3.3.2 Experimental Procedures

Separation and Quantitation of PCR Products Derived from Competitive PCR Reactions, e.g. for Detection of MRP-mRNA.

Materials:

1) PCR amplified MRP-DNA fragments (see chapter 3.4)
2) pBR322-Hae III-digest (Sigma)

3) *TSK DEAE-NPR column (TosoHaas): 4.6 mm ID, Length: 35 mm*
   Stationary phase:
   - *Hydrophilic DEAE linked anion exchanger.*
   - *Capacity: > 0.15 meq/ml*
   - *Particle size: 2.5 μm diameter*
   - *pH-range: 2 to 12*
   - *pK_a of anionic groups: 11.2*
4) *Guard column (Perkin-Elmer): 4.6 mm ID, Length: 5 mm*
5) *HPLC-system (Jasco):*
   - *Model PU-980 Intelligent HPLC pump with Low Pressure Gradient Former and Degasser*
   - *Model UV-975 UV/VIS Detector*
   - *Model AS-950 Intelligent sampler*
6) *Mobile phase:*
   - *Buffer A: 25 mmol/l Tris/HCl, 1 mol/l NaCl; pH 9.0,*
   - *Buffer B: 25 mmol/l Tris/HCl; pH 7.0*

Protocol 1

Each quantitation experiment should begin with a run of pBR322-Hae III-digest-standard to prove the column separation quality and reproducibility (Figure 3.3-1a).

a

MRP5\SIGMA1   Start: 22. Aug. 94   08:48

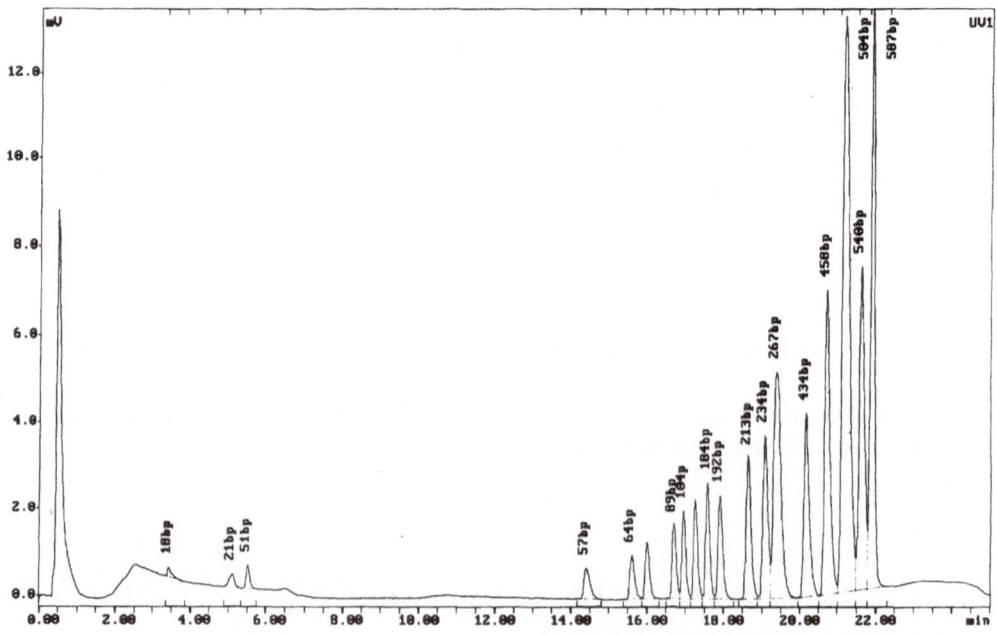

Fig. 3.3-1

b

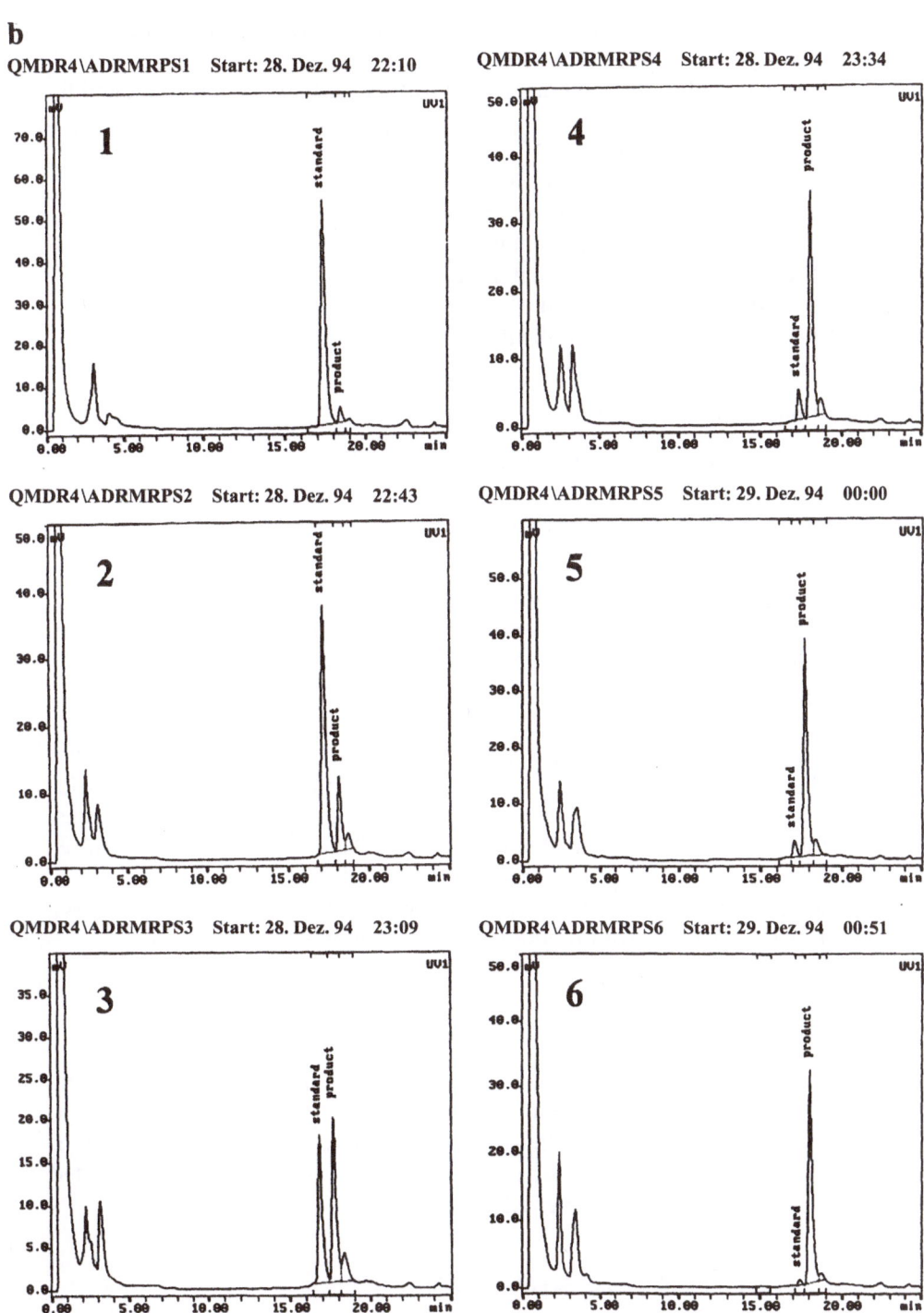

Fig. 3.3-1. Quantitation of DNA by anion-exchange HPLC. a HPLC separation of 20 µl of pBR322 DNA-HaeIII digest (50 mg/ml, Sigma Chemicals). b Detection of amplified MRP fragments. cDNA derived from 100 ng of total RNA isolated from the CCRF ADR5000 cell line was coamplified with decreasing amounts of added competitor fragment. (1) 250 fg, (2) 25 fg, (3) 12.5 fg, (4) 5 fg, (5) 2.5 fg, (6) 1 fg standard per tube were used

Following equilibration 20 µl aliquots of PCR reaction mixes (see chapter 3.4.) were applied to the column.

The mobile phase was of binary composition consisting of buffers A and B.

The following gradient program was employed:

1. Equilibration of the column with 25 % A in B
2. 25 – 45 % A in B: linear gradient up to 0.5 min
3. 45 – 50 % A in B: linear gradient up to 4.5 min
4. 50 – 62 % A in B: linear gradient up to 15 min
5. 62 – 100 % A in B: linear gradient up to 22 min
6. 100 – 25 % A in B: linear gradient up to 24 min
7. 25 % A in B: Equilibration

HPLC operating conditions:

- Operative Pressure: 80 – 100 bar (maximum back pressure: 200 bar)
- Flow rate: 1 ml/min
- Temperature: Room temperature
- UV-detection: 260 nm

Separation time of each run: 25 min

The program used is able to separate DNA-fragments between ~ 20 – 2000 bp. According to the gradient used, DNA fragments differing by approx. 3 bp can be distinguished (Figure 3.3-1 a).

Note: Using PCR samples prepared without a mineral oil overlay and using a guard column, it is unnecessary to extract the reaction mixture for DNA with water-saturated chloroform as recommended in Ref. [3]. After 10 – 15 runs, column cleaning is recommended with 3×50 µl 0.2 mol/l NaOH. Only bidestilled water and sterile filtered buffers should be used.

To get reproducible results use a column oven (with microprocessor controlled Peltier element for optional cooling and heating).

Validation

Estimate the peak integrals (mV * S) by using the respective chromatographic software (e.g. NINA Chromato-Graphic-System, Nuclear Interface GmbH, Münster, Germany). Plot the $\log_{10}$ of product ratio (competitor products to amplified specific DNA) as a function of $\log_{10}$ of amount of the competitor added to the PCR reaction (see chapter 1.1).

3.3.2.1 Benefits of HPLC Analysis of PCR Products

1. Quantitative analysis
   Because of the proportionality between the UV absorption signal and e.g. the concentration of a used reference DNA sample, PCR products can be

G. M. Rothe

Electrophoresis of Enzymes

Laboratory Methods

1994. XII, 307 pp. 59 figs., 58 tabs.
(Springer Lab Manual) Hardcover **DM 98,-**;
öS 764,40; sFr 94,50   ISBN 3-540-58114-6

This book compiles facts and methods on enzyme electrophoresis widely dispersed in hundreds of publications. The author summarizes them in clearly readable tables, in many carefully worked out electrophoresis and more than 140 staining protocols. This book is a "must" for every enzyme laboratory. It will supply the practitioner with profound information on state-of-the-art enzyme electrophoresis.

From the contents: Extraction of Enzymes from Tissues, Cells, and Cell-Organelles. - Methods for Separating Native Enzymes (Electrophoresis in Cellulose Acatate, Starch, Polyacrylamide, Non-denaturing 2D-Electrophoresis). - Sodium Dodecyclsulphate Electrophoresis. - Chemistry of Enzyme Visualization. - A Compilation of Protocols to Visualize Enzymes. - Data Evaluation in Population Genetics and Evolution.

Prices are subject to change. In EU countries the local VAT is effective.

■ ■ ■ ■ ■ ■ ■ ■ ■ ■ ■

Springer

Springer-Verlag, Postfach 31 13 40, D-10643 Berlin,   tm.BA95.08.25
Fax 0 30 / 82 07 - 3 01 / 4 48, e-mail: orders@springer.de

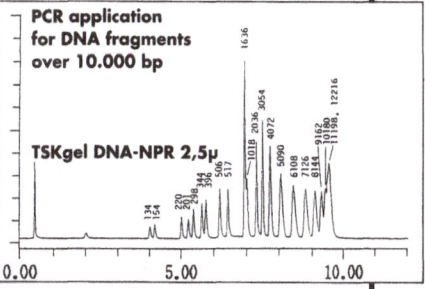

determined according to their peak integrals. Thus nucleic acids can be quantified by HPLC without loss or further specific detection methods. As examples, the quantitation can be performed for amplified HIV-1 DNA, HBV DNA or after RT-reaction for viral RNA and cellular mRNA, respectively, from biological lysates.

2. Monitoring
HPLC can be used to follow up both selection of optimal PCR conditions and reaction kinetics so that the exponential phase of the reaction can be estimated because this is a crucial factor in most PCR quantitation strategies. The monitoring is performed by subjecting aliquots of a PCR sample amplified for different cycles to HPLC analysis. In this way, the number of PCR cycles where the target DNA is exponentially amplified can be distinguished from the point where the reaction reaches the plateau phase (see chapter 1.1).

3. Preparative HPLC
HPLC can be used for the purification of larger DNA aliquots (e.g. primers). Chromatographically separated DNA can be collected using a fraction collector followed by simple precipitation of DNA. The isolated DNA can be concentrated and desalted further by dialysis or additional column chromatography for further applications. Purified DNA may serve as a target for sequencing reactions or in the case of biotinylated or digoxigenin-labeled probes for hybridization or in the case of small oligonucleotides for capillary electrophoresis and further separation with higher resolution.

4. High yield without loss
Because separation and quantitation are performed in a one step process no substantial loss of or any change in the DNA determined occurs. These advantages in the determination and quantitation of DNA makes HPLC suitable as a reference method in relation to other proofs and quantitation methods for nucleic acids (e.g. gel electrophoresis, Northern or Southern blotting, silver and ethidium bromide staining, autoradiography).

5. Determination without processing of DNA
Quantitative DNA-detection by HPLC is performed by UV-absorbance at 260 nm and requires no further processing of samples (e.g. hybridization with labeled probes). Product yield can be determined from a calibration curve generated by separation of commercially available DNA mass standards (e.g. distributed by Gibco).

6. Specificity, sensitivity and precision
Artifacts (e.g. nonspecific PCR products) are recognized immediately according to the well-defined retention times of analyzed DNA mainly depending on charge and therefore on the predicted molecular weights. The detection limit is mainly determined by the sensitivity of the UV-VIS detector and is about 0.2 to 0.4 ng of DNA per peak. Thus UV detection of DNA is much more sensitive than staining of DNA bands within agarose gels after electrophoresis by ethidium bromide but nearly as sensitive as silver staining or conventional hybridization techniques. Higher sensitivity may be achieved with a laser fluorescence measurement after addition of intercalating fluorescence dyes to samples containing DNA (e.g. Hoechst 33258,

Hoechst AG or YO-PRO-1, Molecular Probes, Eugene, USA) before the HPLC run is started [2,6].

The precision of the HPLC assay (for intra- and interassay estimations) is about 4% [2].

3.3.2.2 Drawbacks of HPLC Analysis

1. Time consuming technique
   Whereas 48 to 96 samples may be amplified by one PCR run in 1–2 hours (a similar number of samples can be analyzed by electrophoresis, blotting or the ELOSA technique, respectively) with HPLC one sample only can be analyzed per run. Without an auto-sampler, the analysis capacity is limited to about 15–20 runs a day.
2. Expensiveness
   Because of the high costs of HPLC apparatus, materials and working time HPLC analyses are much more expensive then electrophoresis, blotting, autoradiographic or immunochemical methods.
3. Limited life of columns
   The life expectancy for anion exchange columns is limited to a maximum of 400–500 runs. This is due to low pore size of the column material and some waste may accumulate onto the column leading to poor results (e.g. asymmetric peaks). To prevent early aging of the column, the guard column should be changed after approx. 100 runs.

Note: A quantitative DNA-determination can be performed alternatively with *capillary electrophoresis* (CE) especially for separation and analysis of primers or short PCR products (20–300 bp).

CE separations occur in a narrow capillary tube (50–100 mm ID) filled with buffer containing 1% hydroxyethyl cellulose or polyacrylamide. Separation of DNA is performed by an entanglement process in a single-step voltage gradient (150–400 V/cm). A linear relationship exists between molecular weight and migration time.

To the DNA sample a fluorescent intercalating dye is always added for laser-induced fluorescence detection at 520 nm in a Beckman P/ACE 2050 capillary electrophoresis instrument with a Laser module 488 argon ion laser. Data are recorded on Waters Millennium 2010 software (Millipore, Bedford, USA) by comparing size and separation in relation to internal standards used in each run located mostly on both sites of the oligonucleotides of interest. To overcome some drawbacks of separation of oligonucleotides by HPLC with CE-analysis, single base resolution can be obtained on 60-cm columns. Using shorter capillaries a 3 bp resolution in the 100–400 bp region is obtained within 10 min [5]. The highest precision and resolution of small oligonucleotides has been achieved with constant voltage separations at 170 V/cm and run times of 20 min approximately. Like HPLC the CE separation can also be used for quantitation of PCR products [6].

Trouble Shooting

For long-time storage, the HPLC-column must be protected from the growth of microorganisms by incorporating a bacteriostat such as 0.05% (w/v) NaN_3 into buffer B or 20% acetonitrile in water. Column problems due to clogging will result in increased column back pressure. Partial clogging of the frit can result in tailing peaks due to uneven sample distribution. Simple backflushing with half of the normal flow rate is often successful in cleaning the top frit. Additionally one should take a look at the trouble shooting guides on the data sheets distributed with the column.

References

1. Henninger H-P, Hofmann R, Grewe M, Schulze-Specking A, Decker K (1993) Purification and quantitative analysis of nucleic acids by anion-exchange high-performance liquid chromatography. Biol Chem Hoppe-Seyle; 374:625–634
2. Katz ED, Bloch W, Wages J (1992) HPLC Quantitation and identification of DNA amplified by the polymerase chain reaction. Amplifications; 8:10–13
3. Gaus H, Lipford GB, Wagner H, Heeg K (1993) Quantitative analysis of lymphokine-mRNA expression by a nonradioactive method using PCR and anion exchange chromatography. J Immunol Methods; 158:229–236
4. De Kant E, Rochlitz CF, Herrmann R (1994) Gene expression analysis by a competitive and differential PCR with antisense competitors. BioTechniques; 17:934–942
5. Butler JM, McCord BR, Jung JM, Allen RO (1994) Rapid Analysis of the Short Tandem Repeat HUMTH01 by Capillary Electrophoresis. BioTechniques; 17:1062–1070
6. Butler JM, McCord BR, Jung JM, Wilson MR, Budowle B, Allen RO (1994) Quantitation of PCR products by capillary electrophoresis using laser fluorescence. J Chromatogr; 658:271–280

Chapter 3.4

Quantitation of Absolute Numbers of mRNA Copies in a cDNA Sample by Competitive PCR

Th. Köhler

3.4.1 Theoretical Background

In this chapter a competitive PCR assay using an internal DNA standard is described. Competitive PCR is a quantitative adaptation of the PCR method in which a known number of copies of an exogenous synthesized and added DNA (or RNA) is amplified together with the sample in the same PCR tube (see chapter 1.1).

The full procedure starting with standard generation, standard calibration and finally performing competitive PCR and detection of synthesized DNA as a requirement to measure absolute mRNA levels is described as an example for the determination of MRP gene expression. However, the technology described here may be adapted to the measurement of any other mRNA of interest.

MRP gene overexpression was shown to be associated with complications occurring with chemotherapy of tumors. The successful treatment of tumors with cytostatics is often hindered by primary or acquired resistance of the tumor cells to cytotoxic drugs. Besides the well known mechanism mediated by expression/overexpression of the MDR-1 gene (see chapter 2.3), recently a distantly related protein called "multidrug resistance-associated protein" (MRP) was described which belongs to the same ATP-binding-cassette superfamily [1] and that was shown analogously to confer multidrug resistance [1–3]. Therefore it may be of great clinical interest to find out if MRP expression could be responsible for the multidrug resistance phenomenon especially in those tumors resistant to any therapy and lacking MDR-1 expression.

Whereas MDR-1 is expressed in only a few cell types the MRP gene product was reported to be ubiquitously found [1]. To detect gene overexpression in small amounts of tissue e.g. found in bone marrow taps or tumor biopsies qualitative PCR strategies fail to provide sufficient information. Here quantitative PCR methods clearly offer an essential advance.

As a precondition to performing competitive PCR analysis, a simple and easy to mimic method to generate suitable competitor fragments by site-directed mutagenesis (see chapter 1.2) is described in this section. Standard calibration and measurement of the in-vitro synthesized PCR products were performed by anion-exchange high performance liquid chromatography (HPLC, see chapter 3.3). This highly sensitive method was shown to allow exact standard and PCR product quantitation devoid of non-specific PCR products and contamination derived from purification procedures.

The competitive PCR method was used to determine absolute levels of MRP mRNA in the T-lymphoblastoid cell line CCRF CEM and the drug resistant, MDR-1 expressing variants CCRF ADR5000 and CCRF VCR1000.

3.4.2 Experimental Procedures

- Generation of a dsDNA standard for competitive PCR
- Purification and calibration of the standard DNA fragment by HPLC
- Quantitation of the absolute number of MRP-mRNA copies by competitive PCR in cDNAs prepared from the two drug-resistant cell lines CCRF ADR5000 and CCRF VCR1000 and the parental line CCRF CEM

3.4.2.1 Generation of an Internal DNA Standard for Competitive PCR

All primers used were designed and checked by using automated analysis software as described in chapter 1.2. The competitor fragment was generated by a simple site-directed mutagenesis protocol [4, 5]. A DNA fragment suitable as internal standard adapted for a chosen PCR product may be generated by a simple one-step technique using the (+) standard PCR primer and a linker primer carrying the original (–) primer sequence on its 5′ end (see Figure 1.2-2, chapter 1.2). This short procedure was shown to be sufficient to generate suitable standards in a single day [4] if the linker primer used is suitably designed and does not form loops or other secondary structures with itself.

Materials

1) cDNA from the human drug-resistant cell line CCRF ADR5000
2) Primers:
- *(+) primer MRP3 (nt 751–770) 5′ GCTCGTCTTG TCCTGTTTCT 3′*
- *(–) primer MRP4 (nt 1071–1090) 5′ CTCCACCTCC TCATTCGCAT 3′*
- *linker primer MRP5 (nt 1071–1090/967–986)*
- *5′ CTCCACCTCC TCATTCGCAT CCTTCTTCCA GTTCTTTACC 3′*
3) Reagents and equipment necessary to perform a standard PCR reaction (see chapter 2.3, Protocol 1 and section 3.4.2.3)

Protocol 1: Generation of a MRP Competitor DNA Fragment

1. Prepare 6 PCR microfuge tubes, each containing:
   - 4 μl cDNA
   - 2 μl primer MRP3 (about 100 ng per μl)
   - 4 μl MRP5 fusion primer (about 100 ng per μl)
   - 8 μl dNTP-mix (dUTP)

- 5 µl 10 PCR buffer (Perkin-Elmer)
- 2 µl UDG (1 : 10 diluted with H_2O)
- 22 µl H_2O
2. After initial 10 min denaturation at 94 °C and cooling down to 72 °C add 3 µl AmpliTaq polymerase (Perkin-Elmer) according to the "Hot-Start" protocol (see chapter 2.3)
3. Amplify for 40 cycles using following amplification conditions:
   - 94 °C, 0:30 min
   - 53 °C, 0:30 min
   - 72 °C, 1:00 min
   - final step: 72 °C, 10:00 min
   - followed by cooling to 4 °C

Note: Because the fusion primer used for standard generation is twice as large as the (+) standard primer it must be introduced in 2-fold higher quantity into the PCR reaction mixture to ensure adequate primer concentrations preventing asymmetric amplification.

Fragments containing dU nucleotides may be used as internal competitors as well as native DNA with the intention of guaranteeing contamination-free work. However, despite this certainty, all steps involving standard storage, dilution and addition to PCR tube should be performed in a separate room distinct from the room where the PCR reaction takes place.

Specificity of the Competitor Obtained

Test the amplified competitor DNA sequence by subjecting aliquots of the PCR sample containing the standard (256 bp) fragment to further reamplification e.g. with the original primers (MRP3 and MRP4) as shown in Figure 3.4-1. The

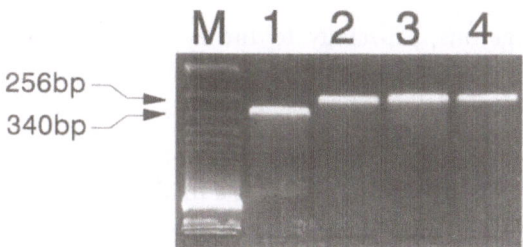

Fig. 3.4-1. Generation of a competitive MRP-standard according to Celi [4] and Förster [5]: In each PCR reaction the same 5′ primer (MRP3) was used. Marker (*M*): 123 bp ladder (Gibco), *Lane 1:* 340 bp fragment amplified with the original (-)-primer (MRP4) and cDNA from the CCRF ADR5000 cell line, *Lane 2:* 256 bp fragment amplified with the internal 3′-linker primer (MRP5) and ADR cDNA as template, *Lane 3:* 256 bp fragment generated with the 3′-linker primer and the 340 bp product as template, *Lane 4:* 256 bp fragment amplified with both original primers and purified competitor as template

synthetic DNA fragment is demonstrated as exhibiting the desired properties: 84 bp shorter in length than the endogenous derived fragment but carrying the same primer binding sites as the target sequence.

Additional evidence for the competitor specificity may be obtained by conventional restriction analysis as described in chapter 2.3 or from a calibration curve generated by HPLC fractionation of a molecular weight standard when plotting the bp versus the retention times.

3.4.2.2. Purification and Calibration of the Standard Oligonucleotide

The synthesized standard fragment may be purified by agarose gel electrophoresis followed by cutting out the ethidium bromide stained bands and DNA extraction from the agarose slices. Alternatively a pure competitor product may be yielded by quantitative HPLC. However, because of the high NaCl-concentration necessary to eluate the fragment from the ion-exchange resin and the disadvantageous pH value (about 9.0) the DNA recovery following ethanol precipitation is very poor. Therefore the gel purification should be strongly favored. The recovery by this method is about 1–4 µg DNA from 6 PCR reactions sufficient to perform thousands of competitive PCR reactions.

Protocol 2: Purification and Calibration of the MRP Competitor

1. Subject the pooled PCR mixtures containing the amplified MRP standard fragment (Protocol 1) to electrophoresis through a 1.5% agarose gel (7 × 8 cm), pre-stained with ethidium bromide (see section 2.3.2.2). Use combs forming slots of sufficient capacity. Perform electrophoresis at 100 V for 45 min.
2. Place the ethidium bromide stained gel on a transilluminator and excise the DNA bands respective to the 256 bp fragment.

>>**Ultraviolet radiation is dangerous, especially to the eyes. To minimize exposure make sure that the UV source is adequately shielded and the face is protected by a safety mask that blocks UV light efficiently.**<<

3. Purify the DNA from the gel slices as described in section 2.5.2.2. When eluting the DNA from the resin with 50 µl sterile water a final concentration of about 10–20 ng/µl (roughly examined by UV spectroscopy) can be yielded. Store aliquoted at −20 °C.
4. Accurate standard calibration using HPLC [6]:
   Mix 4 µl DNA mass ladder (Gibco) containing 5, 10, 20, 40, 60 and 100 ng per 2 µl in the 100, 200, 400, 800, 1200 and 2000 bp bands, respectively, with 36 µl buffer B (see chapter 3.3.) and apply 20 µl of the mixture on a TSK DEAE-NPR column (TosoHaas) in duplicate. Employ a discontinuous gradient program and proceed as described in chapter 3.3.

Generate a standard curve as shown in Figure 3.4-2. Chromatography in parallel an aliquot of purified standard and quantify the standard as accurate as possible. Make sure that the standard concentration is exactly in the range determined by the standard curve.

Trouble Shooting

The main prerequisite for the detection of absolute mRNA copies in a given sample is the exactly known concentration of the initially added standard. Quantify the used PCR standard as accurately as possible at least in duplicate. Separation of both mass ladder and purified standard fragment must be performed under identical conditions (i.e. at least on one day using the same charge of eluent). The standard should be diluted and introduced into the PCR sample in an appropriate volume (about 2–5 µl) to avoid pipetting inaccuracies.

3.4.2.3 Quantitation of MRP mRNA by Competitive PCR

In order to roughly estimate the expression of the sought after mRNA species in the sample, a titration assay has to be initially performed. Titration means that aliquots of cDNA prepared from total or mRNA have to be mixed with various dilutions of the internal standard. Following amplification the resulting PCR products are measured and the 1:1 molar ratio of target and competitor is roughly assessed. Depending upon the results of the initial titration assay the number of dilution series has to be adapted to the particular expression level of the target mRNA. A dilution series in 1:2 to 1:5 steps is recommended.

Materials:

1) *cDNA samples (1 µg RNA per 20 µl-reaction mix)*
   cDNA from the human drug-resistant cell line CCRF ADR5000
   cDNA from the human drug-resistant cell line CCRF VCR1000
   cDNA from the parental line CCRF CEM
2) *the MRP mastermix sufficient for one PCR reaction (50 µl) contains:*
   * *2 µl primer MRP3 (about 100 ng per µl)*
   * *2 µl primer MRP4 (about 100 ng per µl)*
   * *8 µl dNTP-mix*
   * *5 µl 10 × PCR buffer (Perkin-Elmer)*
3) *Taq polymerase (Perkin-Elmer), freshly diluted to 0.5 U per µl with H_2O*
4) *Standard dilutions (1 amol = 10^{-18} mol ≅ 171 fg of the MRP standard):*
   Standard (1): 1.46 amol (250 fg); (2): 0.15 amol (25 fg); (3): 7.31×10^{-2} amol (12.5 fg); (4): 2.93×10^{-2} amol (5 fg); (5): 1.46×10^{-2} amol (2.5 fg); (6): 9.75×10^{-3} amol (1.7 fg); (7): 5.85×10^{-3} amol (1.0 fg); (8): 2.93×10^{-3} amol (0.5 fg); (9): 1.46×10^{-3} amol (0.25 fg); (10) 9.75×10^{-3} amol (0.12 fg)

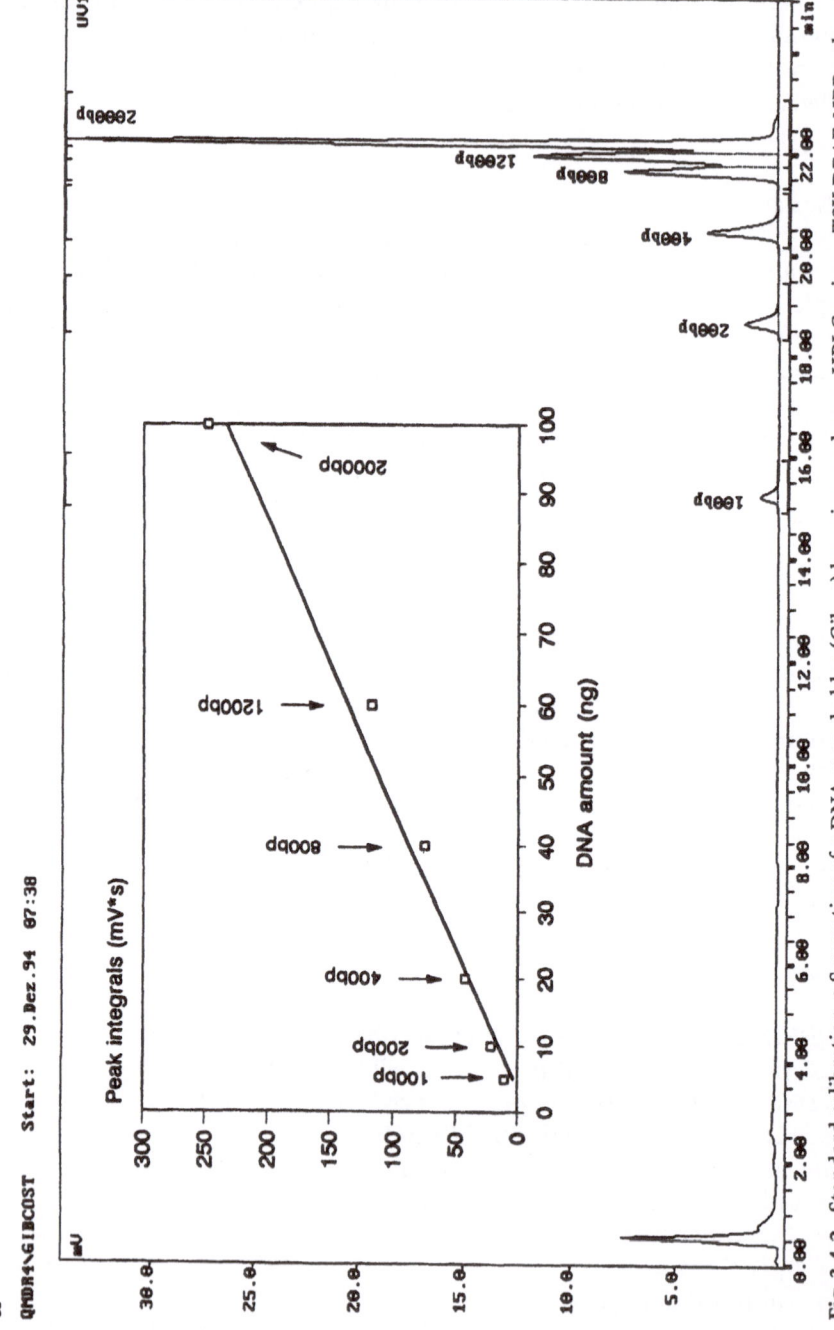

Fig. 3.4-2. Standard calibration. a Separation of a DNA mass ladder (Gibco) by anion-exchange HPLC using a TSK DEAE-NPR column (TosoHaas) to generate a calibration curve.

b   MRP\MRPST2   Start: 18.Aug.94   17:09

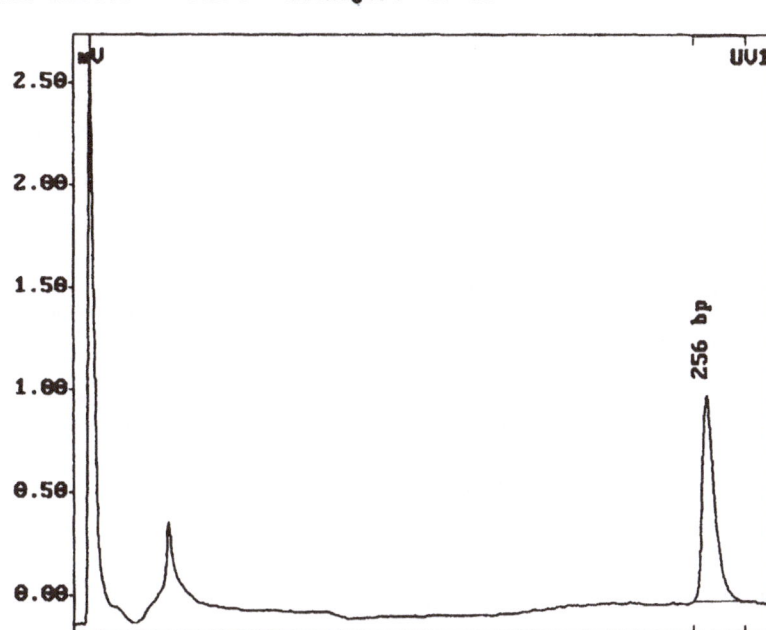

Fig. 3.4-2. b Separation of the gel-purified 256 bp MRP standard fragment performed by the same procedure. The resulting peak integral was compared with the calibration curve to calculate the accurate concentration

5) *Uracil-DNA glycosylase (Boehringer, Mannheim), freshly diluted to 0.1 U per μl, see note!*
6) *MicroAmp reaction tubes (Perkin-Elmer)*
7) *GeneAmp 9600 thermal cycler (Perkin-Elmer)*

Note: The nature of PCR makes it susceptible to contamination problems. Contamination with internal PCR standards leads to irreproducible results and continuous underestimation of the measured RNA transcripts. General substitution of dTTP with dUTP in generation of standards and PCR products derived from both target and DNA standards in conjunction with the use of uracil-DNA glycosylase (UDG) has been demonstrated to selectively prevent carryover contamination.

Protocol 3: Competitive PCR

1. Preparation of the reaction mixture: for maximum reproducibility of results mix the following solutions (sufficient for 8 amplification reactions):
   • 136 μl mastermix (MRP)
   • 16 μl diluted UDG
   • 16 μl cDNA (derived from a standard RT-reaction, see chapter 2.2)

Table 3.4-1. Pipetting scheme for competitive PCR

tube number	1 μl	2 μl	3 μl	4 μl	5 μl	6 μl	7 μl	8 μl
H_2O	24	24	24	24	24	24	24	30
reaction mixture	21	21	21	21	21	21	21	–
mastermix	–	–	–	–	–	–	–	17

- after a brief centrifugation place the tubes in the thermal cycler and perform UDG-sterilization for 15 min at 37 °C followed by inactivation of the enzyme by incubation at 94 °C for 10 min, >>do not add the standards prior to UDG-sterilization!<<

- perform one PCR cycle (see below) to synthesize the starting ds DNA template, remove the tubes from the cycler and add the standards at room temperature, usually in a separate room

standard dilution 1 [4]	2	–	–	–	–	–	–	–
standard dilution 2 (5)	–	2	–	–	–	–	–	–
standard dilution 3 (6)	–	–	2	–	–	–	–	–
standard dilution 4 (7)	–	–	–	2	–	–	–	–
standard dilution 5 (8)	–	–	–	–	2	–	–	–
standard dilution 6 (9)	–	–	–	–	–	2	–	–
standard dilution 7 (10)	–	–	–	–	–	–	2	–

- centrifuge briefly and place the tubes again in the 72 °C hot cycler (>>be careful, prevent burning!<<), add Taq-polymerase and run the desired cycle program

Taq-polymerase	3	3	3	3	3	3	3	3

2. Proceed further as shown in Table 3.4-1
3. Temperature profile of amplification: Run 30–40 cycles (i. e. work in the plateau phase of reaction!) under the following conditions (Cycle parameters):
   - 94 °C, 0:30 min
   - 53 °C, 0:30 min
   - 72 °C, 1:00 min
   - final step: 72 °C, 10:00 min followed by cooling to 4 °C

Validation of Competitive PCR Analysis to Determine Absolute Levels of mRNA

After the amplification is completed remove 10-μl-portions from each sample and pipette into separate tubes. Add 1 μl of nondenaturing bromphenol/xylene cyanol loading buffer (see appendix) mix and centrifuge briefly. Resolve the

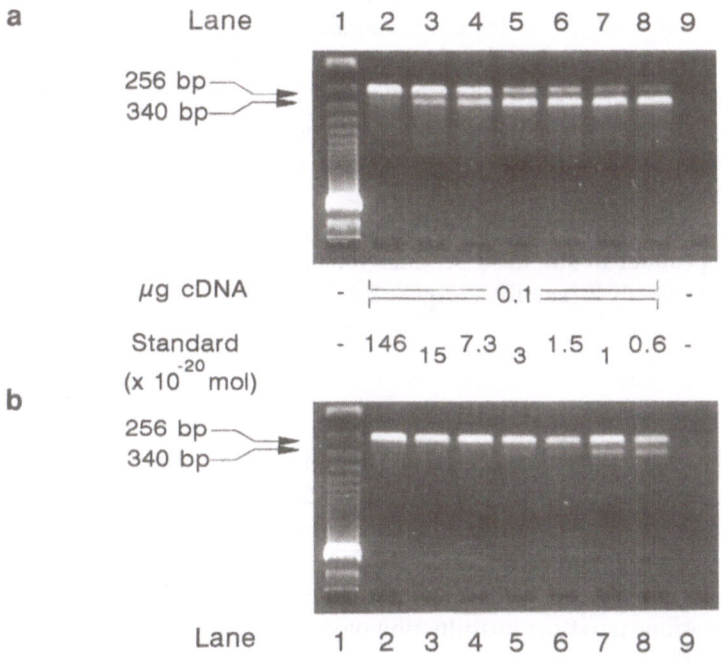

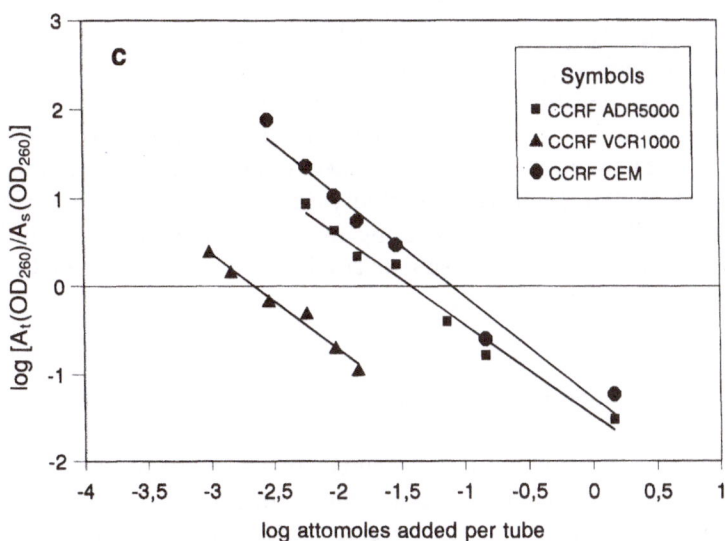

Fig. 3.4-3. Quantitation of MRP mRNA for the drug resistant T-lymphoblastoid cell lines CCRF ADR5000 and CCRF VCR1000 compared to the parental line CCRF CEM. Simultaneous amplification of MRP cDNA derived from the ADR **a** and VCR **b** variant at different concentrations of internal standard. **c** HPLC analysis of the results of a competitive PCR experiment under plateau conditions (40 cycles). The Log of the ratio of amplified target to competitor product is graphed as a function of the Log of a known amount of competitor added to the PCR reaction. When the molar ratio of target and competitor is equal to 1 the absolute target concentration can be read from the X axis

mixture on a 2% agarose gel as described in chapter 2.3. Following electrophoresis through the ethidium bromide stained gel and transillumination, calculate the relative amounts of PCR-fragments corresponding to the internal standard (256 bp) and the endogenous product (340 bp) by quantitative densitometry using e.g. the ImageMaster evaluation software (Pharmacia Biotech) or ScanPack (Biometra) in combination with a CCD-camera, or apply HPLC analysis (see chapter 3.3).

Calculate the absolute amounts of target product by plotting the Log of the ratios of target product to standard product versus the initial amounts of the standard added to the PCR reactions (Figure 3.4-3).

Trouble Shooting

If dsDNA standards are used make sure that the competitive reaction starts with an endogenous template that is also double stranded. Therefore perform one PCR cycle before adding the standard. If you do not so, the template can be underestimated in the range of about 100% depending on the cycle number when the PCR process is stopped. An alternative method which does not require this pre-amplification step is the introduction of single-stranded standards into the competitive PCR reaction mixtures as recommended and performed by de Kant et al. [7].

In our experience, the main limitation in performing competitive PCR is the correct quality and quantity of the introduced competitors. Most suppliers advise using only freshly prepared standard solutions because in solutions containing the lowest quantities of the standard, the possibility of unspecific adsorption to the microfuge tube rises dramatically. However, when constant remaining quantities of an unrelated DNA are added to the standard dilution we could find an substantial improved durability in solution. Therefore we recommend testing, following repeated freezing and thawing, whether the same standard dilution series yields equal results over a long period when using the same cDNA as the template.

3.4.2.4 Sensitivity and Reproducibility of the Assay

The sensitivity of competitive RT-PCR is extremely high allowing the detection of at least 0.0002 MRP mRNA copies per cell without any problems. Employing this technique we have been able to observe obviously good intraassay coefficients between 2–6% (determined from 3–6 separate amplifications of GAPDH and collagenase 1 mRNA, respectively) and interassay coefficients of ±2–9% (from 3 separate GAPDH amplification experiments) using confectioned batches of standard dilutions.

References

1. Cole SPC, Bhardway G, Gerlach JH, Mackie JE, Grant CE, Almquist KC et al. (1992) Over-expression of a transporter gene in a multidrug-resistant human lung cancer cell line. Science; 258:1650–1653
2. Grant CE, Valdimarsson G, Hipfner DR, Almquist KC, Cole SPC, Deeley RG (1994) Over-expression of multidrug resistance-associated protein (MRP) increases resistance to natural product drugs. Cancer Res; 54:357–361
3. Hamaguchi K, Godwin AK, Yakushiji M, O'Dwyer PJ, Ozols RF, Hamilton TC (1993) Cross-resistance to diverse drugs is associated with primary cisplatin resistance in ovarian cancer cell lines. Cancer Res; 53:5225–5232
4. Celi FS, Zenilman ME, Shuldiner AR (1993) A rapid and versatile method to synthesize internal standards for competitive PCR. Nucl Acids Res; 21:1047
5. Förster E. An improved general method to generate internal standards for competitive PCR. BioTechniques; 16:18–20
6. Köhler T, Laßner D, Rost A-K, Leiblein S, Remke H (in press) Polymerase chain reaction related approaches to quantitate absolute levels of mRNA coding for the multidrug resistance-associated protein and P-glycoprotein. In: Proceedings of the 2nd International Symposium "Drug resistance in Leukemia and Lymphoma", March 6–8, 1995 Amsterdam, "Advances in Blood Disorders" series, Harwood Academic Publishers
7. De Kant E, Rochlitz CF, Herrmann R (1994) Gene expression analysis by a competitive and differential PCR with antisense competitors. BioTechniques; 17:934–942

Acknowledgement

We thank Drs. Volker Gekeler, Byk-Gulden GmbH, Konstanz, and Heyke Diddens, Medical Laser Center, Lübeck, for providing us with the cell lines CCRF CEM, CCRF ADR5000 and CCRF VCR1000.

Appendix

Buffers and Solutions:

- DEPC-H_2O:
  Treat 1 liter of deionized H_2O with 1 ml DEPC at room temperature under constant shaking for at least 1 hour and autoclave.
- dNTP mixture (Promega, Pharmacia):
  12.5 μl dATP, dCTP, dGTP, dTTP (dUTP) from a 100 mmol/l stock solution, diluted to 1000 μl with H_2O
- Kanamycin:
  Dissolve 50 μg kanamycin /ml LB agar in corresponding volume of DEPC-treated water. For sterilization of the solution press through a sterile-filter tip using a gauge (0.2 μm pore size) directly by mixing with the warm agar medium and distribute the agar in sterile Petri dishes (10 cm diameter, 20 ml agar solution per dish).
- LB medium (liquid medium):
  10 g tryptone, 5 g NaCl, 5 g Bacto Yeast extract (DIFCO), add H_2O dest. to 1 liter
  Prepare the medium and autoclave the solution immediately.
- LB medium (solid medium, for agar plates):
  10 g tryptone, 5 g NaCl, 5 g yeast extract (DIFCO), 10 g Bacto-Agar (DIFCO), add H_2O dest. to 1 liter
  Prepare the medium and autoclave solution immediately. The agar can be poured into the Petri dishes after autoclaving. Prepare agar containing antibiotics as mentioned in chapter 1.3.
  Cool down the medium (to about 45 °C) and add the appropriate antibiotic (kanamycin or ampicillin) if necessary.
- MOPS-buffer, 10 × conc.:
  Dissolve 20.6 g MOPS in 400 ml DEPC-treated H_2O; adjust to pH 7.0 (20 °C) with NaOH; add 8.4 ml 3 mol/l Na-acetate and 20 ml 250 mmol/l EDTA, pH 8.0 (20 °C); bring to final volume of 500 ml with DEPC-H_2O and sterilize the solution by filtration through a 0.2-μm filter; 10 × MOPS solution is stored at room temperature, protect from light!)
- Na_2HPO_4-solution (1 mol/l)
  The stock solution is composed of 179 g $Na_2HPO_4 \times 12$ H_2O and 4 ml 85 % H_3PO_4 per liter, pH 7.2 (20 °C).
- PCR buffer, 10 × conc. (Perkin-Elmer):
  100 mmol/l Tris/HCl, 500 mmol/l KCl, 15 mmol/l $MgCl_2$, 0.01 % (w/v) gelatin; autoclaved; pH 8.3 (25 °C)
- Phosphate-buffered saline (PBS):
  4.3 mmol/l Na_2HPO_4 * 7 H_2O, 1.4 mmol/l KH_2PO_4 137 mmol/l NaCl, 2.7 mmol/l KCl
- Potassium acetate solution:
  Per 200 ml potassium acetate solution mix 120 ml of a 5 mol/l K-acetate solution, 23 ml glacial acetic acid and 57 ml H_2O, adjust to pH 4.8.
- Sample loading buffer (4 × conc.):
  0.25 % (w/v) bromphenol blue, 0.25 % (w/v) xylene cyanol, 30 % (v/v) glycerol in H_2O)
- SSC buffer, 20 × conc.:
  per 1 litre 175.3 g NaCl, 88.2 g sodium citrate; pH 7.0

- TAE electrophoresis buffer (50 × conc. stock solution):
  242 g Tris/HCl, 57.1 ml acetic acid, 37.2 g EDTA ad 1000 ml H$_2$O; pH 7.5
- TBE buffer, 10 × conc.:
  108 g Tris; 55 g boric acid add 1000 ml aqua bidest, 0.2 mol/l EDTA
- TE-buffer:
  10 mmol/l Tris/HCl, 1 mmol/l EDTA; pH 8.0 (20 °C)

Suppliers of chemicals

- Amersham
  USA: Amersham North America, 2636 South Clearbrook Drive, Arlington Heights, IL 60005.
  Tel.: 800-323-9750, Fax: 800-228-8735
  Europe: Amersham International, Amersham place, Little Chalfont, Buckinghamshire HP7 9NA, U.K.
  Tel.: 01494-54400, Fax: 01494-542266
- Angewandte Gentechnologie Systeme (AGS)
  Europe: Angewandte Gentechnologie Systeme GmbH, Rischerstrasse 12, D-69123 Heidelberg, Germany.
  Tel.: 06221-831023, Fax: 06221-840610
- AT Biochem
  USA: 30 Spring Mill Drive, Malvern, PA 19355.
  Tel.: 610-889-9300, Fax: 610-889-9304
- Boehringer Mannheim GmbH
  USA: Boehringer Mannheim Biochemicals, P.O. Box 50414, Indianapolis, IN 46250.
  Tel.: 800-262-1640, Fax: 317-576-2754
  Europe: Boehringer Mannheim GmbH, Sandhofer Strasse 116, P.O. Box 310120, D-68298 Mannheim, Germany.
  Tel.: 0621-759-0, Fax: 0621-759-8509
- Clontech Laboratories
  USA: Clontech Laboratories Inc., 4030 Fabian Way, Palo Alto, CA 94303-4607.
  Tel.: 415-424-8222, Fax: 800-424-1350
  Europe: ITC Biotechnology GmbH, P.O.B. 103026, D-69020 Heidelberg, Germany.
  Tel.: 06221-303907, Fax: 06221-303511
- Difco
  USA: Difco Laboratories, P.O. Box 331058, Detroit, MI 48232-7058.
  Tel.: 313-462-8500, Fax: 313-462-8517
  Europe: Difco Laboratories GmbH, Ulmer Straße 160 a, Postfach 101486 D-86004 Augsburg, Germany.
  Tel.: 0821-443391, Fax: 0821-443891
- DuPont
  USA: DuPont Company, 549-3 Albany Street, Boston, MA 02118.
  Tel.: 800-551-2121
  Europe: DuPont de Nemours GmbH, Diagnostics & Biotechnology, DuPont Strasse 1, D-61343 Bad Homburg, Germany.
  Tel.: 06172-872600, Fax: 06172-872540
- Dynal
  USA: Dynal Inc., 475 Northern Boulevard, Great Neck, NY 11021.
  Tel.: 800-638-9416, Fax: 516-829-0045
  Europe: Dynal International, P.O. Box 158 Skoyen, N-0212 Oslo, Norway.
  Tel.: 2-529450, Fax: 2-507015
- Eppendorf Gerätebau Netheler and Hinz GmbH
  Europe: P.O. Box 650670, D-2000 Hamburg 65, Germany.
  Tel.: 040-53801-0, Fax: 040-53801556
- Fluka Chemie AG
  Europe: Messerschmittstrasse 17, D-89231 Neu-Ulm, Germany.
  Tel.: 0731-973-03, Fax: 0731-973-3160

- Greiner Labortechnik
  Europe: Greiner GmbH, Maybachstraße 2, D-72636 Frickenhausen, Germany.
  Tel.: 07022-501-0, Fax: 07022-501-514
- Hoefer Scientific Instruments
  USA: 654 Minnesota Street, Box 77387, San Francisco, CA 94107-0387.
  Tel.: 800-227-4750, Fax: 415-821-1081
  Europe: Pharmacia Biotech Europe GmbH, Munzinger Strasse 9, D-79111 Freiburg,
  Germany.
  Tel.: 0761-4903193, Fax: 0761-4903246
- Hoffmann-La Roche AG
  Europe: Hoffmann La-Roche AG, Roche Diagnostika, D-79639 Grenzach-Wyhlen,
  Germany
  Tel.: 07624-14-0, Fax: 07624-2576
- Hybaid
  USA: National Labnet Co., 650 Hadley Road, South Plainfield, New Jersey 07080.
  Tel.: 291-283-4555, Fax: 201-561-5634
  Europe: Waldegrave Road 111–113, Teddington Middlesex, TW11 8LL, U.K.
  Tel.: 0181-977-3266, Fax: 0181-977-0170
- Invitrogen Corporation
  USA: Invitrogen Corporation, 3985 B Sorrento Valley Blvd., San Diego, CA 92121.
  Tel.: 800-955-6288, Fax: 619-597-6201
  Europe: Invitrogen BV De Schelp 26, 9351 NV Leek, Netherlands.
  Tel.: 05945-15175, Fax: 05945-15312
- ITC Biotechnology
  USA: Clontech Laboratories Inc., 4030 Fabian Way, Palo Alto, CA 94303.
  Tel.: 415-424-8222, Fax: 415-424-1064
  Europe: ITC Biotechnology GmbH, P.O.Box 103026, D-69020 Heidelberg, Germany.
  Tel.: 06221-303907, Fax: 06221-303511
- Jasco Labor und Datentechnik GmbH
  USA: Jasco Inc., 8649 Commerce Drive, Easton, MD 21601-9903.
  Tel.: 800-333-5272, Fax: 410-822-7526
  Europe: Robert-Bosch-Strasse 11, D-64823 Gross-Umstadt, Germany.
  Tel.: 06078-74949, Fax: 06078-74006
- Life Technologies/BRL
  USA: Life Technologies/BRL Inc., 8400 Helgermann Court, P.O.Box 6009, Gaithersburg, MD
  20884-9980
  Tel.: 301-840-8000
  Europe: Life Technologies/BRL GmbH, P.O. Box 1212, D-76339 Eggenstein,Germany.
  Tel.: 0721-780444, Fax: 0721-780499
- MBI Fermentas
  Europe: Fermentas Molecular Biology Instruments, Graiciuno 8, Vilnius 2028, Lithuania.
  Tel.: 0122-641279, Fax: 0122-643436
- MWG Biotech
  Europe: Anzinger Strasse 7, D-8017 Ebersberg, Germany.
  Tel.: 08092-24071, Fax: 08092-21084
- National Biosciences (NBI)
  USA: National Biosciences Inc., 3650 Annapolis Lane, Plymouth, MN 55447-5434.
  Tel.: 800-747-4362, Fax: 800-369-5118
  Europe: MedProbe, Postboks 2640, St. Hanshaugen, N-0131 Oslo, Norway.
  Tel.: 2-2200137, Fax: 2-2200189
- Perkin-Elmer Corporation
  USA: Perkin-Elmer Cetus, Main Avenue, Norwalk, CT 06856.
  Tel.: 800-762-4001, Fax: 203-761-2542
  Europe: Perkin Elmer Holding GmbH, European Sales Support Center, Bahnhofstrasse 30,
  D-85588 Vaterstetten, Germany.
  Tel.: 08106-381-115, Fax: 08106-6697

- Pharmacia Biotech
  USA: Pharmacia Biotech, 800 Centenial Avenue, Piscataway, NJ 08854.
  Tel.: 800-526-3593, Fax: 800-329-3593
  Europe: Pharmacia Biotech AB, Björngatan 30, S-75182, Uppsala, Sweden.
  Tel.: 18165000, Fax: 18143820
- Promega
  USA: Promega Corporation, 2800 Woods Hollow Road, Madison, WI, 53711.
  Tel.: 800-356-9526, Fax: 608-277-2516
  Europe: Serva, Carl-Benz-Strasse 7, P.O.B. 105260, D-69042 Heidelberg, Germany.
  Tel.: 06221-502-0, Fax: 06221-502-188
- Qiagen
  USA: Qiagen Inc., 9600 De Soto Avenue, Chatsworth, CA 91311.
  Tel.: 800-426-8157, Fax: 800-718-2056
  Europe: Qiagen GmbH, Max-Volmer-Straße 4, D-40724 Hilden, Germany.
  Tel.: 02103-892230, Fax: 02103-892233
- Roth
  Europe: Carl Roth GmbH + Co., Schoemperlenstraße 1–5, Postfach 211162,
  D-76185 Karlsruhe, Germany.
  Tel.: 0721-560 60, Fax: 0721-560649
- Schleicher & Schüll
  USA: Schleicher & Schuell Inc., 10 Optical Avenue, Keene, NH 03431.
  Tel.: 603-352-3810, Fax: 603-357-3627
  Europe: Schleicher & Schuell GmbH, P.O. Box 4, D-37586 Dassel, Germany.
  Tel.: 05561-791-0, Fax: 05564-2309
- Serva Feinbiochemika
  Europe: Serva Feinbiochemika GmbH, & Co. KG, P.O. Box 105260, D-69042 Heidelberg,
  Germany.
  Tel.: 06221-502-0, Fax: 06221-502188
- Sigma
  USA: Sigma Chemical Company, P.O. Box 14508, St. Louis, MO 63178-9916.
  Tel.: 800-325-3010, Fax: 800-325-5052
  Europe: Sigma-Aldrich Chemie GmbH, Grünwalder Weg 30, D-82041 Deisenhofen, Germany.
  Tel.: 0130-5155, Fax: 0130-6490
- Sorin Biomedica
  Europe: 13040 Saluggia (Vc), Italy
  Tel.: 0161-4871, Fax: 0161-487545
- Stratagene
  USA: 11099 North Torrey Pines Road, La Jolla, CA 92037.
  Tel.: 800-424-5444
  Europe: Stratagene GmbH, P.O.Box 105466, D-69044 Heidelberg, Germany.
  Tel.: 06221-400634, Fax: 06221-400639
- TosoHaas
  USA: 156 Keystone Drive, Montgomeryville, PA 18936.
  Tel.: 215-283-5000, Fax: 215-283-9385
  Europe: TosoHaas GmbH, Zettachring 6, D-70567 Stuttgart, Germany.
  Tel.: 0711-13257-0, Fax: 0711-13257-89
- United States Biochemical (USB)
  USA: United States Biochemical Corporation, P.O. Box 22400, Cleveland, OH 44122.
  Tel.: 216-765-5000, Fax: 216-464-5075
  Europe: United States Biochemical GmbH, Niederstedter Weg II, D-61348 Bad Homburg,
  Germany.
  Tel.: 0130-855085, Fax: 0130-304755
- Whatman
  Europe: Whatman Scientific Ltd., Whatman House, St. Leonard's Road, Maidstone, Kent,
  ME16 OKS, U.K.
  Tel.: 01622-6766-70, Fax: 01622-677-011

Subject Index